THE OCTOPUS IN
THE PARKING GARAGE

THE OCTOPUS IN THE PARKING GARAGE

A CALL FOR CLIMATE RESILIENCE

rob verchick

COLUMBIA UNIVERSITY PRESS

New York

Columbia University Press
Publishers Since 1893
New York   Chichester, West Sussex

Copyright © 2023 Rob Verchick
Paperback edition, 2025
All rights reserved

Library of Congress Cataloging-in-Publication Data

Names: Verchick, Robert R. M., author.
Title: The octopus in the parking garage : a call for climate resilience /
Rob Verchick.
Description: New York : Columbia University Press, [2023] | Includes
bibliographical references and index.
Identifiers: LCCN 2022035934 (print) | LCCN 2022035935 (ebook) |
ISBN 9780231203548 (hardback) | ISBN 9780231219013 (paperback) |
ISBN 9780231555104 (ebook)
Subjects: LCSH: Human beings—Effect of climate on. | Adaptation (Biology) |
Natural disasters—Risk assessment. | Resilience (Ecology)
Classification: LCC GF71 .V47 2023 (print) | LCC GF71 (ebook) |
DDC 304.2/5—dc23/eng20221103

Cover design: Henry Sene Yee
Cover images: Shutterstock

FOR MADI

There is no end
To what a living world
Will demand of you.

—Octavia Butler

Contents

THE OCTOPUS IN
THE PARKING GARAGE

PART I

Understanding Resilience

chapter 1
Let's Talk About the Octopus

I n the parking garage of the Mirador 1000 condominium complex—days after a bronze "supermoon" had swallowed Miami's night sky—there were two signs that things had not yet returned to normal. One was the green pool of water that sloshed around the tires of about a dozen stranded cars. The other was the tentacles.

The source of the pool was a fountain of water bubbling loudly from a corroded floor drain. Its eddies shimmered in the building's artificial light. A few feet away, protruding just above the water's surface, was the splayed form of an octopus, its rubbery limbs heaving against the cool cement. On the morning of November 14, 2016, when Miami resident Richard Conlin happened upon this sight, he did what any of us would have done. He shot photos and video with his phone and posted them on Facebook.[1] The "Octopus in the Parking Garage" went viral.

"OMG! It really IS an octopus," commented one follower on social media.

"Was it alive?" asked another, "Did it get back into the bay?"

"What? Did you grill it!?" came a third, attaching a sly-faced emoji.

And from a fourth: "How does an octopus get into a garage?"

As it turns out, the octopus *was* real and very much alive. Happily, it was spared the grill and probably did survive. Conlin reported in an updated post (later verified by the *Miami Herald*) that members of the building's security staff had quickly filled a bucket with saltwater and transported the creature back to sea. "They believe it got back safely," he wrote.

How the octopus got there is a longer tale of quirky plumbing and climate breakdown. The Mirador 1000 sits near the ocean, along with the parking facility and its accompanying drainage pipes. Those pipes, which feed runoff into the ocean, were built years ago and were originally suspended above Biscayne Bay.

But sea levels are rising on account of climate change, and this, along with other human impacts on the bay, now leave many drainage pipes submerged and vulnerable to flooding at high tide.

In the case of the Mirador 1000, a cycle of high tides, known as a "king tide," was already in gear and had been further amplified by the unusually strong gravitational pull of the moon, which was at a point in its orbit most proximate to Earth, an event called a "supermoon." The "supermoon" and the "king tide"—in an environment already altered by sea level rise—caused the storm drain to reverse and burp up a cephalopod. As Kathleen Sullivan Sealey, a marine biologist at the University of Miami, told *Yale Climate Connections*, "the octopus is just putting a kind of a personal face on a big problem."[2]

I won't argue with the "personal face." Once you Google Conlin's octopus pics, you won't unsee them. But the octopus in the parking garage is a broader symbol too. It's an eight-armed alarm bell, a reminder of how ubiquitous climate impacts have become. It's an urgent call for recognizing and preparing for the many climate disruptions to come. That kind of preparation is what I'll refer to as *climate resilience*, which includes a range of things we might do to promote safety and security in an era of unprecedented ecological risk. But before we go there, we have to talk about this octopus (figure 1.1). Because if we can't keep sea life out of our parking spots, what else can't we do?

To put that question in context, I arranged a meeting with Sealey, the Miami biologist whose interview first introduced me to this story. Sealey specializes in coastal restoration and recently coauthored a book on climate-change risk and Miami's real estate markets.[3]

The Mirador 1000, Sealey told me, was one of several luxury residences located in Miami's South Beach neighborhood, an area known for its ocean views and astronomical rents. From the photos, Sealey said the octopus appeared to be of the Caribbean reef variety, *Octopus briareus*. "It was a pretty good size," she said. Indeed, the length of its muscular mantle (the headlike structure containing all the organs) looked to be about eighteen inches. The creature's arms were somewhere about twice that length.

I asked her what the octopus was doing in a submerged drainage pipe. Probably hunting, she told me. These pipes attract a lot of sea life because they flush a lot of fertilizers from lawns and cropland into the bay, which boosts algal growth. She said an experienced scuba diver can sometimes identify a submerged pipe just by the "little halo of algae" surrounding its opening. Octopuses

1.1 The octopus in the parking garage, as shown on Richard Conlin's Facebook feed.
(Photo by Richard Conlin)

also just like to hide. The drain outfalls "make big deep dark holes," according to Sealey. "Octopus love them." As for flooded parking facilities, not so much. "Of course, the octopus didn't *want* to go into the parking lot. That was not where it wanted to go. It was *forced* up there."

The kind of tidal flooding that forced our friend into reserved parking is forcing a lot of change in Southeast Florida. That's because these "sunny day floods," as the locals call them, have been increasing because of sea level rise. Today the region experiences ten tidal floods per year, a number that has risen significantly over the last decade. By 2045, scientists think Southeast Florida will see 240 floods per year.[4] The damage is dramatic and widespread. Seawater spouts from storm drains; manholes blow their tops; major roads are swamped with several inches—or feet—of polluted water. The floods snarl traffic, maroon pedestrians, kill lawns, destroy property. Even worse, as tidal pressure builds below the surface, saltwater is squeezed into the aquifers that provide drinking water to six million South Floridians.[5]

Agencies and communities in the region are responding on many fronts. The city of Miami Beach alone is raising its streets by several feet and revamping its sewer system. It has plans to install an advanced pumping system, along with several emergency generators in case of power outages. Southeast Florida's water district is being forced to consider a series of massive engineering projects to protect municipal water supplies from saltwater intrusion. It's a good start, but there's a long marshy road ahead. Home sales have begun falling precipitously in the flood zones. Gazing into the future, property values in much of the region seem to be mamboing on a bubble.

When I asked Sealey if Miami-Dade County had made any progress in preventing the kind of drainage back-ups that affected the Mirador 1000, she told me the work was frustratingly slow. Some engineers have suggested modifying the county's drainage pipes with one-way valves, which allow water to flow out but not in. "But that's ridiculous," she said, "because the edges of so many of these old pipes are chipped." Sealey shook her head and sighed. "I haven't really seen any kind of effective valve." Plus, it's not like anyone has mapped the hundreds of pipes lying beneath Biscayne Bay. Despite the occasional algal halo, outfall drains are really hard to find.

Meanwhile, garage flooding persists. "Everybody's Porsche and Mercedes sits in seawater, which is *not* a good thing," Sealey emphasized. "I think now there is actually a lease surcharge if your address is in certain zip codes because you're going to drive that car through saltwater, which is *terrible* for it."

Those in the Porsche-Mercedes set have it easy. As is typical with catastrophic harms, it's the people with less money and less political power who will bear the brunt of Florida's climate disaster. Their communities are often located in more geographically vulnerable areas with limited municipal budgets. For them, seawalls, storm pumps, and elevated highways will be financially out of reach. I asked about Miami-Dade's Office of Resilience, which was founded, in part, to deal with climate-change impacts. "Their whole office is property values, property values, property values," she explained. "Because if they don't have people paying their taxes, they don't have money to build more resilient structures."

It's a tale of two Atlantises. "Some people's houses will be *behind* the seawalls and some people's houses will be *in front of* the seawalls. If you are in your seventies and you've got millions of dollars in a hedge fund, go live on Star Island," Sealey advised, referring to an exclusive residential island in Biscayne Bay (of which, by the way, the Mirador 1000 has a fabulous view). "If you lose that home, you'll be fine. But if you've got three or four kids, and you're trying to build equity in your home, and you've got credit card debt, you need to move."

━━━●━━━

At this point we've all heard the dire warnings: our climate system is quickly unraveling. The breakdown can be traced to a primary human cause: the release of greenhouse gases, most notably, carbon dioxide emitted from burning fossil fuels.[6] These gases accumulate in the atmosphere like an invisible blanket, trapping heat and warming the planet. There's nothing wrong with heat-trapping gases per se. Carbon dioxide occurs naturally in the atmosphere and plays a critical role in regulating Earth's temperature. Without naturally occurring carbon dioxide, the average temperature would just slightly exceed a winter in Fairbanks, Alaska. Before we started burning fossil fuels, the natural concentration of carbon dioxide in the atmosphere was about 280 parts per million.[7] Now it's over 414 parts per million, which is nearly 48 percent higher.[8]

The extra energy getting trapped in the atmosphere shows up in a lot of places. Some of it is melting glaciers and ice caps. Some of it is evaporating water in one locale and dumping it faraway in another. When the evaporated water in the air condenses, a corresponding release of energy adds power to hurricanes and other storms. Throughout the process, average temperatures of oceans and land masses rise.

Thus the world we inhabit—the blue dot that holds all life we know—is getting hotter, drier, wetter, and weirder. Since the late nineteenth century, the dot's average surface temperature has risen by 1 degree C (1.8 degrees F). By 2100, that temperature could register up to 4.2 degrees above preindustrial levels on the Celsius scale.[9] (That might not sound like much, but according to many studies, an increase of 2 degrees C would be enough to obliterate some island nations and destroy whole ecosystems.) Our coastlands are sinking. Forests are bursting into flame. Droughts and heat waves are getting worse. Rain patterns are seesawing in the United States and in many other parts of the world. We hear this from a consensus of the world's climate scientists, from farmers whose families have plowed land for generations, from U.S. federal courts and judicial tribunals abroad. We hear this from the pope.[10] And we know that, absent some ambitious intervention, the climate's Great Unraveling will accelerate.

On top of this comes another dose of hard reality: As a matter of physics, global warming cannot be reversed very quickly. Scientists attribute this to something called "thermal inertia." Imagine a pot filled with water on a gas stove. At around 2,000 degrees C, the flame is twenty times the water's boiling point. But, depending on the amount of liquid, it could take several minutes for the water to boil. Once you extinguish the flame, you must wait even longer—ninety minutes or so—for the boiling water to cool back down to room temperature. Thermal inertia describes the water's delay in absorbing and releasing the flame's heat.

Global warming works the same way. As temperatures rise in the atmosphere, the oceans slowly absorb most of the heat, functioning like a giant hot water bottle. Turn the atmospheric heat down, and the water bottle will release the stored heat, but only gradually. Based on the work of the climatologist James Hansen, scientists believe thermal inertia is responsible for a delay between cause and effect of about forty years.[11] A forty-year "climate lag" implies that average temperatures of the last decade were caused by emissions I contributed to back in 1980, when, at sixteen, I got my first set of wheels—a gas-guzzling Chevy Blazer with a "Big Block" 450 engine. What this means today is that even if we could magically shut off *all* greenhouse-gas emissions tomorrow, we would still be locked into another forty years of climate fluctuation. Of course, we can't shut off emissions tomorrow, which means that under even the rosiest scenarios, the Great Unraveling will be with us for several generations to come.

Another point: Even when the Earth's temperature stabilizes generations after we end human-caused emissions (I'm being optimistic here), all the carbon dioxide we pumped into the sky during our carefree days will still be up there unless we figure out a way to remove it. (On its own, atmospheric carbon takes millennia to disperse and return to the rocks from which it came.) So, even a future "stabilized" climate means one that is hotter and more disruptive than the one that you or I grew up with.

<hr/>

If tales of tentacles have you spooked, you might be tempted to take Kathleen Sealey's advice and just move from whatever risky locale you're in. That's one way to chase resilience. But there are limits. I teach a graduate course called "Environment and Infrastructure," which is about managing nature and the built infrastructure to protect communities from disasters. The whole course, need-less to say, is backlit by global warming.

On the first day, I start with an icebreaker exercise I call "Escape from Cli-mate Change." We all know, I say, that the seas are swelling and temperatures are rising. That's bad news for warm cities on the coast like Los Angeles, Miami, and Norfolk. And then there's New Orleans, where I live. It's true we've got a massive levee, but, well…you never know. So imagine you've lost faith in your city's infrastructure and you want to escape. Where would you go?

Seattle, a student often says. It's rainy and gray now, but in the future you'll be able to grow oranges. Sorry, I say, western Washington is losing its snowpack to warmer weather; the local counties are panicked over water shortages.[12] The coastal rainforests have seen hotter and deadlier wildfires as plant and soil mois-ture dissipates. (And did I mention the region is also overdue for a massive earthquake and tsunami?)[13]

Rapid City, another will say. It's far inland and close to the Sturgis motorcy-cle rally to boot. But as in the Northwest, South Dakota's Black Hill forests are losing moisture too, I say. In the area near Rapid City, the risk of wildfire is expected to climb 600 percent in the next fifty years.[14] You'll need more than a biker's bandana to keep the smoke out of your lungs.

Alaska? Nope. The permafrost is thawing, and more than a dozen coastal vil-lages are sliding toward the sea. And as the frozen soils melt, scientists expect a release of millions of ancient bacteria and viruses to which humans have lost immunity.[15]

Then I tell the class about my friend Mathew Hauer, a sociologist at Florida State University who studies climate migration. In a 2017 paper in *Nature Climate Change*, he estimated that 13 million people displaced by sea level rise in warm coastal cities would move inland to places like Las Vegas, Orlando, and Hous-ton.[16] But you'd be wrong to follow the crowds, I say. Las Vegas, a town forever on the sizzle, is today warming faster than any other American city, breaking records for heat waves and heat-related deaths.[17] Orlando is bracing for more climate-related inland flooding and mosquito-borne illness.[18] As for Houston, the epic floods of Hurricane Harvey are a grim reminder of what that city's unchecked development has in store as warmer water fuels stronger Gulf storms.

This is not to say that there are *no* places that offer relative respite from cli-mate crisis. The mayors of Duluth, Minnesota, and Buffalo, New York, have each considered positioning their cities as refuges in a warming world. But there are not as many safe spots as you might think.[19] "We can't outrun climate change," I tell my students. "We've got to *outsmart* it."

Outsmarting climate change requires resilience. I earlier defined that as "recognizing and preparing for" climate impacts. To expand on that, we might say resilience is the capacity to *manage* and *recover* from a climate impact in a way that preserves a community's *central character*—the parts of its history, culture, and economy that nourish the soul.

———————

Over the last thirty years, most strategies to address climate change did not emphasize resilience and instead focused much more on reducing greenhouse-gas emissions. Curbing emissions is sometimes referred to as "mitigation," because it mitigates, or dampens, the main driver of climate change. It's not that the preparation side was completely unheard of. But in terms of directives and funding, it represented only a tiny part of the overall national and international project.

There were a few reasons for that. Because greenhouse-gas emissions were the cause of the problem, it seemed to many people that it was more important to tackle that problem first. Girding ourselves for the leftover risk could come later. A second reason was that, as compared to climate resilience, the transition from fossil fuels to renewable energy made for a more straightforward conversation. Replacing dirty energy with clean energy turns out to be a complicated problem when you dig into it, but the focus is pretty clear—energy production and use. Climate resilience, on the other hand, involves nearly everything: infrastructure, wildlife management, disaster response, public health, and, more. Squeezing that subject into a manageable size was like trying to fold an octopus into a box.

Finally—and this is a big point—years ago a lot of environmentalists saw climate resilience as a drooping white flag. They thought that talk about elevating homes and expanding drainpipes was admitting defeat or, worse, giving the fossil-fuel industry permission to do nothing and say the risks were manageable. In his 1992 best-seller *Earth in the Balance*, Al Gore even called adaptation a "kind of laziness."[20] (Gore later reversed that characterization, but not until 2013.)[21] For years, resilience efforts were seen as the "poor cousin" of the richer, meatier work of mitigation.[22]

In fact, when I first started presenting my research on climate resilience at academic conferences in the early 2000s, I was often challenged by academics and activists who thought I was copping out. They said I was selling palliative care to a defeated planet. I learned quickly to start my talks with a smile and a disclaimer. "Good morning, everyone," I'd say. "Please get the coffee while it's hot and enjoy the Danish. Now, I think cutting carbon is Job One. I think preparing for climate change is *also* Job One. And if we can't do both, we're hosed."

Fortunately, things have changed. Today climate action takes two distinct priorities: curbing greenhouse gases to fend off worst-case consequences and boosting community resilience to cope with the impacts already mounting. To put it another way, climate action now seeks to *avoid* the harm we can't *manage* and to *manage* the harm we can't *avoid*. The 2015 Paris Agreement on climate change illustrates this twin-engine approach.[23] The global pact aims to limit global temperature rise to no more than 2 degrees C above preindustrial levels while at the same time promising transfers of aid and technology to "strengthen resilience and reduce vulnerability to climate change." Around the world, scores of countries—from Australia to India to Zimbabwe—are drafting plans to prepare for the Great Unraveling.[24]

In 2013 President Obama released a federal "Climate Action Plan," which touted a focused effort to cut carbon. He also ordered federal agencies to "prepare for the impacts of a changing climate that are already being felt across the country."[25] The latter assignment, which involved agency missions ranging from flood insurance to disease control to national parks, was all about "folding the octopus." As a member of the Obama administration, serving at the U.S. Environmental Protection Agency, I helped develop some of the resilience strategies that went into that plan. When President Trump took office, he immediately reversed Obama's commitment to curb emissions and canceled the resilience orders too.

As expected, President Biden quickly reinstated President Obama's earlier plans. Then he shot the moon. In the first two years of his term, President Biden was able to secure passage of two laws that are now driving America's climate agenda beyond what many people had thought was politically possible. Both are booster rockets for climate resilience. The first law, which passed with bipartisan support, is the Infrastructure Investment and Jobs Act of 2021. Among other initiatives, it allocated $47 billion to prepare for a more aggressive climate. The National Oceanic and Atmospheric Administration received half a billion dollars to map and model future flooding. The flood-control budget of the U.S. Army Corps of Engineers was quadrupled. The act allotted a billion dollars for water recycling programs to relieve drought and $250 million for desalination plants to address saltwater intrusion in coastal aquifers. There was funding to help Native American tribes protect their communities from floods and droughts or to relocate.

The second law, the Inflation Reduction Act of 2022, will raise $369 billion over ten years for spending on carbon-free energy and other climate-related priorities. (The legislation also raises money for deficit reduction, health care, and better tax collection.) By investing in more renewable power generation, the act will make the electricity grid not only cleaner but more climate-resilient. The legislation also provides critical new funds to prevent wildfires and survive drought. This is all terrific news and even more reason to read this book. You will want to make sure all this money is being spent wisely and that the communities you care about are being helped by it. Throughout the book I'll point to places where I think money from these laws will make a difference.

But remember, as welcome as this unprecedented investment is, there is still more to do. We started late, and the the song of the climate sirens grows louder every year. Only months before the infrastructure bill passed, the nation's summer had opened with a blaze of wildfires across much of the western United States and ended with one of the largest hurricanes in modern history, knocking out coastal Louisiana and later (as a downgraded storm) drowning communities in New York and New Jersey. The *Washington Post* reckoned that nearly one-third of Americans lived in a county that had been hit by extreme weather during that season, up from about one-tenth during the same period five years ago.[26] As I write, more than 150,000 residents in Jackson, Mississippi, have no running water in their sinks, showers, or toilets because several days of torrential rain and flooding wrecked the city's aging water treatment plant. More than 80 percent of Jackson residents are Black, and roughly a quarter live below the poverty line.[27] There is no national plan to build climate resilience, just as there is no national plan to reduce carbon emissions. Instead, the federal government spends more than $45 billion recovering from disasters, about seven times what it spends on preparing for climate change.[28]

Climate resilience is not about giving up or giving ground. It's about getting real. Today there are good reasons to supercharge our resilience efforts. First, and most important, fighting for resilience is the moral thing to do. There are people *right now* suffering from the Great Unraveling. In the United States and across the globe, they are losing their homes to wildfires and floods; they are losing access to essential fisheries and hunting grounds; they are falling ill during heat waves, power outages, and fever outbreaks. As we will see, a disproportionate share of those who are suffering the most are those with low incomes or people of color. The natural world, too, is on the ropes. We are witnessing *right now* a dizzying scale of extinction not seen since an asteroid crashed into the Gulf of Mexico sixty-five million years ago.[29] Some scientists say one in six species could disappear as the climate heats up over the next century, including nearly all the world's coral.[30] In 2020, bushfires in Australia, aggravated by climate change, tore through more than seventeen million acres of forest, killing tens of thousands of koalas, kangaroos, and other wildlife. An estimated one billion animals may have been lost.[31]

To stand idle in the face of present suffering cannot be defended. When, in 2013, Al Gore reversed his opposition to climate adaption, he did so on ethical grounds, saying he was "wrong in not immediately grasping the moral imperatives" of pursuing carbon reduction and resilience at the same time.[32]

Second, several decades of future climate disruption are already baked into the system. This is the law of thermal inertia. Most of the climate benefits the world will experience from greenhouse-gas reduction will not be felt in our lifetimes, or perhaps even the lifetimes of our children. Caring about future generations means caring about climate resilience.

Third, making our communities more resilient will save a lot of money. Preventing climate-based damage before it occurs is almost always a bargain. Studies, for instance, show that every dollar spent on complying with a better building code saves four dollars of future loss. In the United States, federal grants to help property owners improve storm resilience pay for themselves six times over.[33]

Finally, I believe that climate resilience is the gateway discussion toward broader climate action, including eliminating fossil fuels. As we will see in chapter 5, calls for carbon reduction are incredibly polarizing, politically, and culturally. But conversations about making communities safer from climate-related disaster are less so. In my time at EPA, and now working with state leaders and local communities, I've found that it's easier to excite interest in climate action by focusing on easily observable examples, from a springtime water shortage to a misplaced octopus. We can then ask how a community could avoid such events with more effective reservoirs, retooled drainpipes, or other reforms. Once you get people talking about the values they share, research shows they become amenable to more difficult conversations, like ones involving mining and drilling.

If we are being honest, we must accept that there are also some serious barriers to building resilience at scale. Despite its future payoff, resilience will demand huge financial investment at home and abroad. Resilience efforts also require big investments in time and planning, leading to slowdowns in development and political bickering as citizens get immersed in the details. There's also the point—which we will face throughout this book—that advocates and planners will never know as much as they want to know about science, the engineering, and the economics as they want to know. To act within a cloud of uncertainty is what the climate crisis demands.

Even so, the climate resilience movement is quietly gaining momentum as more communities find themselves directly affected by fires, storms, and other phenomena with a possible climate-change fingerprint. You need only watch the Weather Channel in times of disaster—droughts and wildfire in California, floods in the Midwest and the Gulf Coast, superstorms and "snowpocalypses" in the Northeast. People in those places are already asking, "Did we do that?" The next question to confront is, "How do we survive it?"

This book takes a stab at that question by showing how innovations at the national and local levels can help protect property, ecosystems, and lives. Some innovations I discuss are technological, but the meat of this book involves governance and social cooperation—what experts in the United States and abroad consistently say is most lacking in addressing the effects of climate breakdown. It's also important to understand that in addition to making good personal choices about how and where we live, we also have a duty to make sure our governments and business leaders are up to the task. That means getting involved in local decision making, voting, advocating for smart policies, and discussing climate resilience with colleagues, family, and friends. This book is intended to empower readers to face the climate crisis with courage and to take action that counts. The takeaway here is that grappling with the climate is not a fad, like sous-vide cooking or ax throwing. It is the *new normal*. Like it or not, we will be facing climatic disruptions for decades to come. It's time to talk about the octopus and start that journey.

This book has two parts. Part I explores the meaning of climate resilience, finding examples among early humans on the savannas six million years ago all the way to today's municipal planners in Cedar Rapids, Iowa. The takeaway is that while resilience is surely about technological innovation, it is equally about mindset and *culture*. Shifting the culture toward resilience requires new ways of thinking about our place in the world and a sharper focus on historically marginalized groups—women, people of color, the poor, and others. It's true that people don't care about the climate crisis as much as they should but emphasizing climate resilience could change that by appealing to a broader range of community values. Drawing on new insights from the fields of communications and psychology, part I closes by explaining how that would work.

Part II then surveys the vibrant landscape of climate resilience efforts, in search of instructive stories and rays of hope. My research took me paddling through Louisiana's replumbed bayous, hiking in one of the last refuges of Joshua trees in the Mojave Desert, and scuba diving off Key Largo with citizen scientists working to restore coral reefs. I've met with members of tribal communities in Louisiana and Alaska struggling to relocate from sinking lands. In the last chapter, I offer a list of specific actions that we all can take to build climate resilience, starting in our local communities and progressing to national and international efforts.

Among other things, this is also a story about observation and context. When you see something where you don't expect it to be, it's important to take note and ask why. Kathleen Sealey gives this advice regularly to her students in Miami. "I tell them that in your lifetime, you will have very rapid environmental change," she told me. "You need to start keeping a journal of the environmental change around you. When people say, 'Oh my gosh, I never saw it rain like that before,' you need to write that down. You need to start making yourself aware, because with climate change there are going to be winners and losers, and you don't want to end up being the loser."

All of which got me thinking: What does it *mean* to be a winner? And do there have to be losers at all?

chapter 2
Adapt or Die

The first time I saw her was in a dimly lit hall in the Smithsonian Museum of Natural History. It was noisy and crowded, but neither of us seemed to notice. I stepped forward, nimbly dodging a man pushing a baby stroller. Not taller than the second- and third-graders peering at her through the glass, Lucy leaned from her low perch in a nondescript tree, brimming with expectation, as weightless as a saint. A soft light washed over her face. Lucy's matted hair was parted neatly at the top of the skull in a way that accentuated her narrow brow and high cheekbones. Her eyes wheeled skyward as her left toe stretched onto the sun-bleached earth.

While years have passed, my memory of seeing a life-size model of my three-million-year-old ancestor (*Australopithecus afarensis*), nicknamed "Lucy," lingers on. The Hall of Human Origins, on the first floor of the Museum of Natural History, is still one of my favorite places in Washington, DC. During the Obama administration, I worked as an official at the EPA helping to design climate resilience programs. When I had time on the weekend, I would occasionally slip into this remarkable hall to commune with the family—all 7,500 generations of them (give or take).

The exhibition, which is one of the most popular on the National Mall, immerses you in a journey of human origins, accompanied by stories of survival and extinction in times of climatic upheaval. It's based on decades of research by Smithsonian scientists and collaborating experts from around the world. There are hundreds of casts of prehuman and human fossils illustrating physical and behavioral evolution. I'm fascinated by the life-size tableaus, especially the one of Lucy, that depict scenes in the lives of our early ancestors (with pesky beetles and flies included). You can run your fingers across models of early human skulls to compare the size and shape of braincases and cheekbones over millions of years. You can admire art and jewelry from the Stone Age. Kids really like the "morphing" station, where you can see what you would look like as a protohuman and email yourself the picture.

Here is a saga of resilience and adaptation over the span of some six million years, as our ancestors gradually evolved to walk upright, use tools, and communicate with symbols and language. Their species have names like *Australopithecus afarensis*, *Homo erectus*, and *Homo neanderthalensis*. Scientists refer to them as "hominins" a category that includes the modern human species, all extinct human species, and immediate ancestors like Lucy.

There are many tales of survival in this hall, though all but one ends in extinction. Katherine Sealey's comment about winners and losers (see chapter 1) comes to mind. Today modern humans are top dog. In our 200,000 years of high living, we have accrued an abundance of special advantages, genetic and behavioral. Our species is dubiously dubbed *Homo sapiens*, Latin for "wise man." Half of us, of course, aren't men. And, to be honest, the jury's still out on the wisdom part. I've come to think the better adjective to describe modern humans is "adaptive" because our species is peerless in its ability to improvise and riff on a theme. Still, there may be limits. Change on our planet is careening faster than Lucy or the Neanderthals could have ever imagined. Is there anything we can learn from how they managed? Do their existential trials have anything to say about ours?

In the course of human evolution our ancestors accumulated an impressive list of superpowers that allowed them to thrive in nearly every region they inhabited. That included the ability to walk upright, to make tools, to use language, and much more. What is interesting, as the Hall of Human Origins exhibit points out, is that this roughly six-million-year journey from Lucy's forebears to you and me just happened to coincide with a vast amount of global change, much of it triggered by seesawing atmospheric temperatures.

I learned more about this from Kendra Chritz, a paleoanthropologist at the University of British Columbia who is also a researcher at the Smithsonian's Museum of Natural History. She told me that the temperature shifts described in the Hall of Human Origins were related to an eccentricity in Earth's orbit called the "Milankovitch cycles." What's that? "This is just the earth responding to the pull of other planets in the solar system on it," she told me, pushing and pulling her hands like an accordion player, "which makes it kind of wiggle and move around in its orbit." Over many thousands of years, that wiggling, and the orbital irregularity it produces, "translate to big changes in global climate."

All organisms, Chritz explained, encounter some amount of environmental change, including seasonal variations in the amount of heat, light, and precipitation. But these Milankovitch events unspooled over generations. During that time hominins experienced dramatic shifts in temperature and precipitation that caused an elaborate reshuffling of landscapes, from grasslands and scrublands to woodlands and forests, from cold climates to warmer ones.[1] One way we know this is by studying the oxygen isotopes found in the microscopic skeletons of a kind of amoeba called foraminifera that lived on the ocean floor. Variations in oxygen isotope (the number of neutrons an oxygen atom has) can indicate changes in temperature and glacier distribution over time.

And that wasn't all that hominins had to deal with. While temperatures were doing loop the loops, many regions were rocked by powerful earthquakes and other tectonic shifts. The upward surge of the Tibetan Plateau, to name one example, discombobulated rain cycles in northern China and rejiggered hundreds of miles of placid topography. Seismic jolts displaced lakes and realigned rivers. Volcanic eruptions and forest fires rearranged hunting grounds, watering holes, sources of shelter, and other resources. Unlike seasonal shifts, these events came without warning and sometimes lasted years. Instability and uncertainty were the new normal. For our early ancestors, one rule governed: Adapt or die.

Paleoanthropologists—the scientists who study human evolution—have long admired the hominin talent for adaptation. But they originally discounted the importance of environmental change in its development. For most of the twentieth century, the traditional view was that successful adaption required specialization in an environment free of dramatic change.

According to this view, the most important human adaptations in body structure and behavior, including those responsible for Lucy's upright strolls, were driven by the specific needs of life on the expanding grasslands of eastern Africa. This was called the "savanna hypothesis." But there was a hitch: if key adaptations evolved in response to selection pressures associated with a specific habitat, you would expect those adaptations to be especially fit for that landscape. Thus, you would expect to find hominin species in one kind of environment but not in another. Instead, the fossils of early humans could be found almost everywhere, on the mountain, in the forest, on the plain, and in the rain. Hominins, it turns out, spread like weeds.

Scientists needed a new explanation. That explanation was introduced by a paleoanthropologist named Rick Potts, who is now the founding director of the Smithsonian's Human Origins Program. Potts believed that early humans were not specialized for any particular place. They were instead *generalists* whose impressive quiver of capacities and skills derived from living in a highly variable environment.

"Potts's concept was that the climate is always oscillating," explained Chritz, who once worked with him at the Smithsonian. "This lack of a stable climate resulted in the evolution of our species, which made us very adaptable and able to cope with changes that occur fairly regularly." Potts developed the idea while studying the climatic turnover of animal life in the Olorgesailie region of what is now southern Kenya.[2] He noted that several large mammal species that had previously reigned supreme had gone extinct around 700,000 to 300,000 years ago during a period of repeated environmental upheavals. These species were replaced by animals that tended to be smaller in size and not as specialized in diet or habitat. For example, *Equus oldowayensis*, an ancient zebra of Clydesdale proportions, was a superb surface grazer of that time. But it was eventually replaced by the modern imperial zebra (*Equus grevyi*), which can feed on both grass *and* higher-growing vegetation. The aquatic specialist *Hippopotamus gorgops* similarly gave way to today's slenderized hippopotamus, which is able to amble long distances between watering holes. Sometimes new, more flexible species beat the old-timers at their own game. "There's an extinct short-necked giraffe called *Sivatherium*," Chritz told me, "which used to graze on the plains but eventually died out competing with large-bodied primates who had transitioned to eating grass."

"Evolution," explained Chritz, "is like an ongoing experiment where different things are being tested. And for whatever reason, a particular trait gets selected and if it makes a population more likely to live and reproduce successfully, then that trait will get carried on. So one of the benefits of being a generalist as opposed to a specialist is that as things change, you have a lot more options available for you." Or, as Potts himself recapped in an interview for PBS in 2009, "It's not the survival of the fittest in any one environment but the survival of the more versatile, the more general and flexible creatures that would really persist over time."[3]

Potts's insight, known as the "variability selection hypothesis," is supported by an impressive amount of climatic evidence and has gained a wide following among experts. Still, it is a work in progress. In upcoming years, scientists hope to complete mathematical models that will describe the phenomenon in a fuller and more precise theoretical sense.[4]

But I must say it makes intuitive sense. I grew up in Las Vegas, where almost everyone in my family worked in the local casinos. I learned early on about long odds and the mechanisms of human survival. My grandmother loved to teach me about all the different ways of gambling and how, when you weren't looking, they could eat your lunch.

So imagine you are playing roulette. You have a fixed bankroll, and your goal is to enjoy yourself at the table for as long as possible because you like the free drinks and someone says Beyoncé might possibly stop by. Sizing up your bets, you see two broad options. You can put your money on a small clutch of numbers, hoping for a big win but knowing loss is nearly assured. Or you can "cover the table" by placing a slew of smaller bets to maximize the chance of a payout. For instance, you might place seventeen bets on seventeen pairs of numbers all at once (the "seventeen splits" strategy), which covers all but two pockets the ball could bounce into. The payouts are modest, but you're more likely to win.

What would you do?

Take it from me, when you are looking to survive a night of roulette in a Las Vegas casino, you do *not* want to be a short-necked giraffe with all your chips on a couple of scrublands. You want to be a modern, omnivorous baboon, nimble and fleet, who is just as happy to have her cheeks filled with spiders and snails as with leaves and sedge. Be a generalist. "*Cover the table,*" as my grandmother used say. And if Queen Bey shows up, that's icing on the cake.

I referred to a trove of hominin superpowers in the last section. We are going to unpack them because many of these tools will serve us in adapting to the self-inflicted climate change we face today. But before that, I want to be clear on two points. First, we must recognize that the heating we see today is a rapid *breakdown* of natural systems and in this way is very different from the climatic oscillations of earlier epochs. That was a slow-motion rollercoaster that clattered along for six million years. Our crisis is happening in the blink of a paleoanthropologist's eye. There are lessons that might apply to both periods, but the time frame is very different. Second, the longer time frame made room for genetically based evolutionary changes in body structure and brain mass to accommodate new environments. We don't have time for that, needless to say. We can deploy evolutions in *behavior* just as our ancestors did, making use of the bodies and brains we were born with. That's not nothing. In the right conditions, behavioral change can happen remarkably fast. Indeed, as I hope to make clear in later chapters, our capacity to reimagine our political and social relationships is the key to surviving climate breakdown.

Among the most important capacities, evolutionary scientists cite long-endurance mobility, the capacity to make tools, the use of language and social networking, and, to make all of that work at once, really, really big brains. This portfolio is a coordinated effort between what we might call *genetic* adaption and *behavioral* adaptation.[5] Genetic adaptation, which is caused by random mutations in DNA, might manifest as a change in body structure or biological process and be passed on from generation to the next. A primate species, for example, might develop a squatter, hairier body to survive in colder climates.

You might not think of us humans as adapting in the genetic sense, but in fact we have adapted and still do adapt. Nearly all mammals, for instance, lose the ability to digest breast milk when they reach adulthood. But in many populations, humans retain the ability to digest milk throughout their lives. This phenomenon is a result of natural selection, which favored adult humans who could quaff milk from livestock.[6] Similarly, some mountain-dwelling peoples have the capacity to process oxygen more efficiently than the rest of us, a genetic adaptation that allows them to breath peacefully in the Andes, the Ethiopian Highlands, and the Himalayas.[7]

In contrast, behavioral adaptation describes modifications in behavior and social organizing that are made in response to changing conditions. Donning animal skins and lighting fires under a cold night sky are part of our cultural evolution. So are agriculture and democracy and our global fossil fuel economy. Behavioral adaptation can certainly happen quickly, sometimes within a single generation (checked Facebook in the last hour?). But, more typically, our old ways are "sticky," slowed down or frozen by tradition, ritual, group-based loyalties, or lack of imagination. If everything is going well, that's not a bad thing. But when a crisis approaches, we need to get unstuck. As you might imagine, over long sweeps of time, genetic and cultural adaptation interlock. They are both necessary pieces in what Chritz calls the "evolution puzzle."

On a more recent visit to the Hall of Human Origins, I encounter fewer visitors and am thankful for the opportunity linger and learn. After paying my respects to Lucy in her tree, I'm quickly drawn to an adjacent wall on which a floor-to-ceiling timeline illustrates the oscillations in climate that were occurring during the reign of Lucy's species. I can imagine the gradual withering of the forest floor, the lateral expansion of the grassy plains.

My next stop is Lucy's fossilized skeleton, positioned as if she were on a morning jaunt. Lucy belonged to the genus *Australopithecus*. By about four million years ago, these "Australopiths" had evolved a body structure that allowed them to adjust to changes in moisture and vegetation. Lucy's 3.18-million-year-old skeleton has a humanlike hip bone and knee joints, but with lanky arms. Her grasping fingers are like long spider legs. Her feet are long and look amazingly flexible, which I learn made her both a good walker and skilled climber. This combination gave Lucy's people the ability to go almost anywhere, to scale boulders, roam grasslands, or ascend into the jungle's steamy canopy.

"The Australopiths were really interesting because they had this capability to spend time in trees." Chritz told me. "But it's clear from their skeletal anatomy and morphology that they were spending a lot of time walking around on the ground. Lucy existed in this interesting space of transition." It wasn't an easy life. Even with a versatile body, there were important skills to learn, choices to make.

"You're going to spend a lot of time in the open savanna, you have to deal with differences in thermoregulation—it's pretty hot down there. So you have to deal with how you're going to cool down. And how far are you going to travel? You're on the ground, so you're suddenly prey to all of the large carnivores that existed at that time. You know, it's a lot of different kinds of challenges that were occurring." Lucy's species lived this way for about a million years, which, in case you're wondering, is five times as long as ours has been around.

From her line, the genus *Homo* evolved, and within that, the species *Homo erectus*. These are the oldest known hominins to have possessed bodies proportioned like ours, with stretched-out legs and shortened arms. Living between 1.89 million and 110,000 years ago, *Homo erectus* is also believed to be the first of our ancestors to expand out of Africa, making them our genus's first global carpetbaggers. Fossil discoveries show *Homo erectus* lived not only throughout much of the African continent but also across Asia. Archaeologists have found their buried bones in the sunny grasslands of the Republic of Georgia, in China's chilly Nihewan Basin, and in the tropical marine environment of Java. The species spent all of its time on the ground. But it was otherwise physically versatile and could walk and possibly run very long distances.

Back in the Hall of Human Origins, a few million years away from the Lucy exhibit, I come face-to-face with a nearly complete fossil skeleton of *Homo erectus*. The fossil, known as "Turkana Boy," was discovered near Lake Turkana in Kenya and is estimated to be 1.6 million years old. The boy, though he lived to be only eight or nine, was a Herculean figure, with long, sturdy legs, trim hips, and a barrel chest. A sign behind the glass explains that his lean body was of particular advantage in "hot, dry climates." He stood about five feet tall and would probably have hit six feet had he lived long enough. This guy clearly would have no interest swinging in trees. But I could imagine him cornering boars in the underbrush or hacking the head off a wounded zebra.

I notice a docent wheeling a cart in my direction. She's here to show off some of the museum's hands-on exhibits. Today's topic, she says, is *tools*. I think again of the boar and of the wounded zebra. I want to know more.

The docent tells me that nearly everyone in the *Homo* genus was a wizard with tools. We know hominins even before Lucy had been using stone tools as far back as three million years ago. Tools were an indispensable skill in adapting to the ever-changing environment. Simple tools like flaked or sharpened stones were used to crush and cut food, enhancing the function of hominin's tall, herbivorous teeth. Meat-eating became a thing, wildly expanding the variety of consumable foods. Animals were skinned, flesh was sliced, bones were cracked, and marrow was slurped. This made hominin populations much more flexible and resilient because animal calories were available in virtually any type of habitat an early human might encounter.

Gradually, hominins upped their game, identifying particular kinds of quartz or basalt that could be flaked and sharpened to a razor's edge. This technology, an example of what is called "Oldowan" toolmaking, involved transporting such special stones many miles to inhabited regions where they may have been traded for other things of value.

When *Homo erectus* came on the scene, it began improving on the Oldowan method and by 1.5 million years ago had revolutionized the Stone Age, creating a variety of hand axes with sharper points and straighter blades. There's a famous excavation site near Beijing where *Homo erectus* middens have revealed tens of thousands of fragmentary food bones from birds, turtles, rabbits, fish, pigs, rhinoceros, and deer.[8]

I learn this gradually as we talk, with help from Chritz later on. The docent is a great storyteller and seems to know everything. She smiles and presses a cold, glassy lump into my palm. "This a hand ax," she explains. "It was the Swiss Army knife of the Stone Age." It's shaped like a teardrop, with a rounded end and a small, tapered point. It weighs about two pounds and is very nicely balanced. The edges are wavy and sharp.

The docent says it was made of a volcanic rock called obsidian and can be honed to be sharper than steel. Obsidian blades, she says, are still used in some types of eye surgery. Early humans used hand axes like this one to peel bark, chop tubers, skin animals, and extract bone marrow. She motions toward a life-size display in which two *Homo erecti* are butchering a carcass and fighting off a vulture and jackal trying to steal the meat. I return the hand ax and thank her. Then, after a pause, I ask if she gets many questions about the climate aspect of these displays. "No," she says, "not really."

The obsidian blade is impressive enough. But think for a moment about what else was needed to *make* it and—in a world of circling vultures and jackals—to successfully *use* it. One word: *cooperation*.

Humans and their ancestors were expert networkers and communicators, another of the hominin superpowers. I've already mentioned how the need for specialized rocks to make cutting tools prompted travel and trade among different communities. This not only helped redistribute materials needed for survival, but it also cemented social alliances. By 130,000 years ago, hominins were exchanging materials over distances of more than 150 miles. The resulting relationships may have been critical for survival during times of environmental change where resources in one location grew scarce. Scientists speculate that trade and gift-giving helped set the foundation for other neighborly behavior—the sharing of a watering hole during a drought, for instance, or defending against displaced predators.

Those arrangements required communication, of course. We don't know exactly when spoken language developed among hominins—there are no fossil records of words and grammar. But we do know that hominins communicated through symbols and art as early at least 250,000 years ago. The ability to transmit important and complex ideas in either medium would have made survival in a changing world much easier. Symbolic objects like jewelry and statuary conveyed information about an owner's social status and group membership, which would be helpful in forging alliances. Paintings and drawings showed viewers important aspects of the environment. Symbolic communication allowed early humans to recall the past in a better way, plan for the future, and to build loyalties. As Potts explained in the PBS interview referred to earlier, "There were tremendous survival advantages in being able to create a symbolic world that people in another place could understand. You could begin to have shared beliefs, shared values. And even something as simple as the exchange of valued raw material that could be made into sharp hunting objects . . . meant that these people had a unity, a group unity, which really is the beginning of a modern human way of life."[9]

Long-range mobility. Mastery of tools. A bouquet of social skills. What else did our ancestors need to survive on an erratic planet? A super-sized brain, with which none of the rest would have been possible. For instance, long legs and narrow hips would have helped *Homo erectus* perambulate for miles in search of a food or water, but plotting the journey and minimizing the risks would have demanded unusual creativity and abstract planning skills. Our ancestors' communication systems and complex social interactions, of course, also required strong processing power. And don't forget the hand axes, the shaping of which involved (according to one recent study) a surprising amount of "cognitive control," "information monitoring," and "working memory."[10]

Now, a tantalizing detail: during the first four million years of human evolution (six million to two million years ago), hominin brain size increased very slowly, as analysis of braincase volume in fossilized skulls has revealed. During the next million years, when Lucy's species lived, brain size increased moderately. In the period between 800,000 to 200,000 years, the prehuman brain entered an amazing growth spurt. As it turns out, this growth spurt occurred at a time when Earth's climate was whipsawing between hot and cold more dramatically than it had in any period in the last six million years.[11] That is, the record of human evolution shows a correlation between climate chaos and bigger brains (figure 2.1).

To take the idea one stage further, experts at the Smithsonian have designed a chart showing climate oscillations, from hot to cold, over the last six million

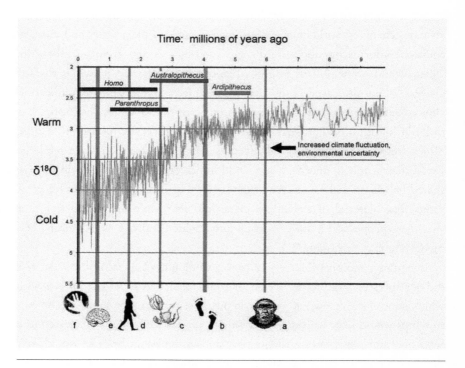

2.1 *Climate fluctuation and human evolution.* This ten-million-year record of oxygen stable isotopes, measured in foraminifera recovered from deep-sea sediment cores, illustrates that global ocean temperature and glacial ice varied widely over the past six million years, the period of human evolution. The $\delta^{18}O$ measurement is the ratio of the heavier ^{18}O isotope and the lighter ^{16}O, which is more easily evaporated from the ocean and sequestered in glacial ice on land. Icons: (*a*) hominin origins, (*b*) habitual bipedality, (*c*) first stone toolmaking and eating meat/marrow from large animals, (*d*) onset of long-endurance mobility, (*e*) onset of rapid brain enlargement, (*f*) expansion of symbolic expression, innovation, and cultural diversity. Data: James Zachos, Mark Pagani, Lisa Sloan, Ellen Thomas, and Katharina Billups, "Trends, Rhythms, and Aberrations in Global Climate 65 Ma to Present," *Science* 292, no. 5517 (2001): 686, http://www.essc.psu.edu/essc_web/seminars/spring2006/jan18/Zachosetal.pdf.

(Courtesy of the Smithsonian Institution)

years. (The information is derived from the oxygen isotopes found in the fora-minifera I told you about earlier.) Over that, they unfurled the evolution of our ancient ancestors, from a genus called *Ardipithecus* to *Australopithecus* to *Homo*. And accompanying that, they plotted the emergence of some of the superpowers we've reviewed—stone toolmaking, long-endurance mobility, rapid brain enlargement, and symbolic expression. Guess what? Nearly all of the major adaptations that scientists associate with human evolution, whether genetic or cultural, are accompanied by the largest oscillations in global climate.[12]

＊

Cover the table. I think about this as I stare into the quiet eyes of a lifelike reconstructed *Homo erectus* head. She is part of a display called "Meet Your Ancestors." Her high cheekbones and square jaw project a knowing confidence. Her eminent nose, broad and convex (perhaps a mechanism for long-distance olfactory navigation), lend her a regal air.[13] On the pedestal below, a sign reads:

> Meet *Homo erectus*
> *Longest-lived member of the human family tree*
> Like others of her species, this female had a larger braincase and more prominent nose than earlier species. *Homo erectus* depended on its large brain to make tools, adjust to new habitats, and care for the old and weak. The species survived for more than nine times longer than our own species has been around.

A generalist of the highest rank, she played the wheel better than anyone. In the end, though, *Homo erectus* eventually disappeared, giving way to Neanderthals in Europe and Western Asia and to us, in Europe and Western Asia and everywhere else. Competition may have been a contributing cause, but more extreme climate change seems to have been the main culprit.[14]

Her assets, it's important to see, are also ours. *Homo erectus* was a highly mobile species, able to retreat from poor environments and resettle in better ones. We've done that. She was also the technological wizard of her time, with the ability to make tools, harness fire, and use medicines. Ditto. *Homo erectus* was also willing to experiment with new systems of social organization geared toward shared survival. We're working on that. A larger brain? We have computers with the computational capacity to model ocean currents, ambient vapor pressure, and atmospheric rivers. Considering all this, Rick Potts has called *Homo sapiens* "a foul-weather species." "Humans," he writes, "do not just withstand change; they adjust to it and make their own modifications. We enclose ourselves for protection and ride out almost any storm."[15]

Still, Potts worries about us. "Our genetic blueprint enables our brains and societies to live creatively in an uncertain world," he writes, "*Homo sapiens* has come into being not by preordained progress, but by trial and delicate survival. . . . As we touch the factors that fueled even our own appearance on Earth, much depends on what we do now."[16]

We know from the history of evolution that we don't have to become the very best at any *single* thing. But we do have to cover our bets and get pretty good at a longer list of tasks. Thinking specifically, this might mean using science and engineering to restore an eroding coast while knowing that at the same time some nearby communities may, in the long run, have to resettle in a more inland community, perhaps sharing real estate or governing power with existing neighbors. At the same time, we must remember that while we have evolved to be expert generalists, many of the millions of the other species that we depend on have not, or their adaptive capacities have been cut short by the environmental upheavals that we have imposed on them.

It is the utmost irony that the extraordinary powers of the human brain have proved to be both our species' salvation on this oscillating world and the potential source of our wholesale undoing. Our enlarged brain was nature's selective response to a changing climate. This gray-celled organ—call it Brain 2.0—was a weapons-grade tool that allowed us to store information, make causal connections, and reimagine the way that we live. It also inspired and empowered us to burn fossil fuel, fill the air with carbon dioxide, and rocket the planet to its boiling point. What's more, that same brain—so famously obsessed with emotional payoff, social loyalties, and the need for control—may pose the most serious barrier to the political and social change needed to outsmart this new, artificial variant of climate change. Is Brain 2.0 any match for Climate 2.0?

I stare again at *Homo erectus*, but this ancestor remains silent. While a college student, I remember reading a novel by Kurt Vonnegut called *Galápagos*. Meant as a kind of tribute to Charles Darwin, the novel is set in the Galápagos Islands after a global financial meltdown. Soon afterward a disease renders Earth's human population infertile, except for a band of shipwrecked passengers stuck on one of the archipelago's remote islands.

Over the next million years, their descendants evolve into a fuzzy new species that looks something like a sea lion—less intelligent than their forebears, but more content. (It's Vonnegut, so stay with me.) The basic message is that human beings, so skilled at engineering their own misfortunes, are ultimately done in by "the oversize human brain." Says the narrator, "The big problem wasn't insanity, but that people's brains were much too big and untruthful to be practical. . . . There was no end to the evil schemes that a thought machine that oversized couldn't imagine and execute."[17]

As a university professor, it's my job to reject that. I've devoted my whole career to the belief that knowledge and thinking are our keys to the kingdom. I believe that. But you can't sulk when the gate is hidden. You can't stop working the problem. During my interview with Kendra Chritz, our conversation at one point turned to teaching. Although our disciplines are different, we both teach courses that involve important aspects of the climate crisis. To engage them, Chritz told me she sometimes asks her students to submit written questions to her before class so she can incorporate the topics into her lectures.

Years ago, when she first started doing this, Chritz said she got a lot of questions challenging the evidence. That doesn't happen much anymore, in part, she believes, because more people are experiencing the real-world effects. Now one of the most frequent questions she receives is one she considers the saddest: "What can we possibly do about this?" "It just seems overwhelming," Chritz told me. "To them it feels like there's nothing we can do."

"What do you say to them?" I asked.

"I say, okay, a 2-degree world is pretty bad," she responds, referring to a scenario in which we are able to limit the rise in global temperature to 2 degrees Celsius above industrial levels. "We're rapidly approaching that, and we're already seeing big changes. But a 2-degree world is so much better than a 4-degree world. And a 4-degree world is so much better than a 6-degree world. In the life of our planet, catastrophes happen, and the planet recovers. So, in the long run the *planet* will be fine. But if we do nothing, there are scenarios that really do threaten *human* survival."

Fretting and doing nothing just aren't options. One thing we must do is cut out the use of greenhouse gases. The other thing we must do is build resilience. Because that sounds big and abstract, we are going to break the job into chunks. We'll meet people who are already doing the work and imagine how it can be brought to scale. We're going to embrace that big brain of ours, understand its strengths and follies, and gradually noodle this problem out.

I'm not saying that adaption is easy. But it's in our DNA.

chapter 3
Sprawling Brains and Rubber Arms

S omewhere in a kelp forest off the coast of South Africa, there's a small, shy octopus with a problem. She is used to dining on mussels and clams, but the day's pickings are slim. Famished and frustrated, her eye settles on an unfamiliar crustacean with markings of coral and black busily combing the seabed. The creature, a species of spiny lobster, has its back turned to our hero. The octopus shambles toward her prey, waggles her arms, then lunges. The lobster zips away, paddling backward with its muscular tail. Over several days, this happens again and again. While the octopus is an expert in corralling fish and sucking clams from the inside out, these lobsters always get away. One day, though, the octopus gets a new idea. She creeps toward the lobster from behind, as usual. But deploying a new technique, the octopus suddenly engulfs the animal in the thin webbing between her arms—bagging the poor creature like a kidnap victim in a gangster film.

This scene unfolds near the midpoint of *My Octopus Teacher*, the Oscar-winning documentary about the relationship forged between a South African filmmaker, Craig Foster, and an octopus living near the coast of his town.[1] Foster, while in the throes of a midlife crisis, follows the octopus for nearly a year, sharing a secret world in which the shy invertebrate reveals herself to be, in the words of one reviewer, "beyond intelligent, dexterous, and resilient."[2] Well, sign me up! The film was recommended to me by Kathleen Sealey, the marine biologist who had helped me make sense of the octopus sighting in the Mirador parking garage.

The scene with the lobster, she said, showed the octopus's ability to adapt to new environments with unfamiliar resources. This was essentially what the Mirador octopus had been doing—perusing the drainpipe for fish and algae—before it got sucked up and dumped into the garage. Foster's octopus friend was luckier. "It went searching shallow waters to look for fish, and instead, it learned to catch lobsters!" Sealey extolled. "That would be like, all of a sudden, figuring out that people deliver pizzas. 'Oh my gosh, here's this big caloric thing I can eat!'" She paused for me to consider the image of a mollusk texting Domino's. "They're very creative."

Indeed, a raft of new research on octopuses, popularized in documentaries like *My Octopus Teacher* and best-selling books like Sy Montgomery's *The Soul of an Octopus: A Surprising Exploration into the Wonder of Consciousness* and Peter Godfrey-Smith's *Other Minds: The Octopus, the Sea, and the Deep Origins of Consciousness*, has revealed this creature to be unbelievably resourceful in negotiating and expanding its habitat. With three hearts, nine brains, and eight semiautonomous limbs, the octopus is an underwater Inspector Gadget. Octopuses use tools, contort through keyhole spaces, and change color to melt into new surroundings. In laboratory settings, they have taught themselves to manipulate light switches (by squirting water at them), recognize human faces, and sneak out of their tanks. Octopuses and other cephalopods can even edit their own genetic information. In 2015, researchers found that common squids have used this trick to acclimatize to various temperature conditions in the ocean.[3]

Given that octopuses are so adaptive and opportunistic, I asked Sealey if there was anything we might learn to boost our own resilience in the face of climate change. "Well, they get a lot of exercise," she said, "and they don't hoard resources." We laughed at that; but it turns out there's even more to learn from these mercurial beings, known for their sprawling brains and rubber arms. I'll turn to that topic shortly.

<hr>

First, let's consider what it means to be "resilient." The word can be traced as far back as the seventeenth century when the philosopher Francis Bacon used it to describe sound waves ricocheting off a wall—what we call an echo.[4] *Resilience* meant literally "to bounce back." A rubber ball is resilient in this way, as is a willow springing back after a burst of wind. However, resilience should be distinguished from "toughness." Toughness is the ability of something to keep going after many blows despite the shape it's in. The surviving car in a demolition derby—all battered and twisted—may not be resilient, but it is tough. We may describe people as resilient, according to the *Oxford English Dictionary*, when they are able to "recover readily" from "a setback, illness, etc." Psychologists have elaborated on this, defining resilience as "the process of adapting well in the face of adversity, trauma, tragedy, threats, or significant sources of stress." If you've ever recovered from a hospital scare, a messy breakup, or a job loss, you can count yourself as resilient in this way. In addition, you may have learned that "as much as resilience involves 'bouncing back,'... it can also involve profound personal growth."[5]

Resilience is also a core concept used by ecologists, whose work, as we will see, contributes to the concept of "climate resilience." Used in the analysis of animal and plant populations, ecological resilience describes "the persistence of relationships within a system and is a measure of the ability of these systems to absorb changes... and still persist."[6] Developed in the 1970s, ecologists deploy resilience measures, bolstered by a strong body of theoretical and mathematical models to help manage complex ecosystems. Since the late 1980s, the concept has expanded to include changes caused by human activity, including climate disruption. Resilience in this sense can be contrasted with what ecologists call "stability"—the tendency to resist change in order to maintain the status quo. A resilient system can fall apart after a major disruption, but it has the ability to reorganize itself in a way that is better adapted to new circumstances. Colonies of spruce budworms—a celebrated example among ecologists—will do this after being decimated by wildfire.[7] (Essentially, these bugs—which are moths with a very hungry caterpillar stage—vary their numbers in multiyear cycles, allowing the forest to repair itself before they devour it again.) When a wave flips my kayak and plunges me into the open sea, I do something similar when I collect my wits, pump the water out of the hull, and awkwardly hoist myself back into the cockpit, more mindful of avoiding such mishaps in the future.

As someone who studies how societies recover from disasters, I'm interested in the resilience of biological and ecological systems; but I focus mostly on the resilience of *human* systems, which requires me to pay special attention to the cross-currents of culture, politics, and law. Since the Stone Age, humans have been adapting to disasters, but it wasn't until the last three hundred years that philosophers and scientists in the West began coiling the skeins of science, sociology, and political theory to weave a pattern that would explain disaster in a fuller context, a pattern that will help us imagine community resilience on a hotter planet.

Once again, we have to turn back the clock, this time to November 1, 1755, for the arrival of what many experts call "the first modern disaster."[8] It wasn't the worst calamity to ever befall humankind, but it was devastating, and the literary correspondence it triggered years later between an ambitious firebrand and an exiled sage would change the course of resilience thinking.

Imagine the morning of All Soul's Day, 1755, in the seaside metropolis of Lisbon, on Portugal's western coast. In darkened churches, the devout light candles to honor lost loved ones. Bells of solace peal from the highest cathedrals. The sea in Lisbon's immense harbor lies flat, flecked with glints of radiant sun. Though free of bustle, we imagine a few crescent-shaped *bocas* wandering the bay. Every so often a net is cast, or a clay octopus trap is dropped into the deep. Then there is an enormous convulsion. Seismic shocks cleave the streets, swallowing buildings and sending stone edifices tumbling. The bay's waters suddenly withdraw, sucked away by an invisible force. Survivors on the dock are stunned to see an exposed plain of mud punctuated by flopping fish, lost cargo, and shipwrecks. When the water returns, it forms a monstrous wave that blasts up the Tagus River—the city's central waterway—and rises so fast that people on horseback are forced to gallop at full speed to escape drowning. Elsewhere, fallen candles in homes and churches start fires that quickly merge into an immense conflagration, devouring neighborhoods and asphyxiating residents even several blocks away. The earthquake's aftershocks continued for three days. For six days, it is said, the ruins of Lisbon blazed, leaving behind "hills and mountains of rubbish still smoking."[9] Historians believe that 85 percent of the city was destroyed and up to seventy thousand people were killed.[10]

In the Age of Reason, the horror of Lisbon was without rationale. For scholars and public intellectuals years afterward, it became something of a philosophical detective story. They mined historical documents, sermons, poems—anything that might suggest *why* something like this would happen and *how* it could be stopped from happening again. Most ordinary Europeans probably believed the earthquake was literally an act of God—punishment for the sins of an extravagant city. Much of the clergy supported this theory, including Gabriel Malagrida, an influential Jesuit priest and member of Lisbon's Royal Court. Some looked for comfort in the Christian writings of Gottfried Wilhelm von Leibniz, whose faith in the central fairness of the cosmos would never be shaken. "A single Caligula" he wrote in 1710, referring to Rome's maniacal third emperor, "has done more evil than an earthquake."[11] Others leaned on Alexander Pope, the famed English poet, who had argued in verse that no natural calamity would ever eclipse God's unfathomable love. "One truth is clear," he insisted, "Whatever is, is RIGHT."[12]

The year of the Lisbon earthquake, Voltaire, the French literary sage, was living in Geneva in exile, having been forced to leave Prussia (and before that France) for his criticism of the Catholic Church. Voltaire had for years politely dismissed the rose-tinted credos of Leibniz and Pope, but after the 1755 quake, he lost his cool. In a blistering poem called "The Lisbon Earthquake," Voltaire railed against any attempt to find justice in fallen towers and "children heaped up mountain high."

> Leibnitz can't tell me from what secret cause
> In a world governed by the wisest laws,
> Lasting disorders, woes that never end
> With our vain pleasures, real sufferings blend;
> Why ill the virtuous with the vicious shares?
> Why neither good nor bad misfortunes spares?
> I can't conceive that "what is, ought to be."[13]

At this point, history might have bid adieu to the aging Voltaire, now growling after a man who'd been dead nearly forty years, were it not for another brilliant literary scold, also living in Geneva—Jean-Jacques Rousseau.

Rousseau, who had just completed his masterwork, *Discourse on Inequality*, had been deeply troubled by Voltaire's poem. He thought the writer was missing something and told him so in a protracted letter of near-Proustian density—a document that would become a classic in the field of disaster studies. While acknowledging the destructive powers of earthquakes and tsunamis, Rousseau insisted that "most of our physical ills are still our own work." "Nature," he reminded Voltaire, "did not construct twenty thousand houses of six to seven stories" in Lisbon's commercial district. "If the inhabitants of this great city had been more equally spread out and more lightly lodged," he continued, "the damage would have been much less and perhaps of no account." Rousseau went on to speculate that the residents' slow evacuation ("because of one wanting to take his clothes, another his papers, another his money") had also contributed to the death toll.

Denouncing the poison of inequality, Rousseau criticized Voltaire for obsessing over fancy citizens of a wealthy city rather than the victims of less sensational disasters. "You might have wished ... that the quake had occurred in the middle of a wilderness," he wrote, "but we do not speak of them, because they do not cause any harm to the Gentlemen of the cities, the only men of whom we take account." Then, rhetorically: "Should it be, that nature ought to be subjected to our laws, and that in order to interdict an earthquake, we have only to build a city there?"[14]

Voltaire never responded to Rousseau's concerns, dismissing their verbal jousts as "amusements." But their conversation and its historical setting illustrate the ways in which educated people thought about disaster. Leibnitz represents the notion of divine justice, or theodicy. When disaster occurs, it should be accepted as either punishment or creative destruction. In this view, science and engineering can do little to mitigate such risks because the underlying forces can be neither understood nor controlled. Theodicy appears to have been the dominant view of large-scale disaster in most civilizations until at least the eighteenth century.

Voltaire rejected theodicy, resigning himself to a universe that was just erratic and heartless. While he celebrated the power of reason, Voltaire did not expect to find it steering the cosmos. At most, he believed, science and technology could help human beings build temporary refuge on an otherwise disorderly planet. "We must cultivate our garden," says Voltaire's hero near the end of *Candide*.[15] But gardens have gates; and beyond them, the woods are just as frightening as before.

To these notions, Rousseau offered his own take, which the sociologist Russell Dynes has called "the first truly social scientific view of disaster."[16] By insisting that "most of our physical ills are still our own work," Rousseau anticipates today's hazard-mitigation experts by centuries. Note Rousseau's attention to the city's design and to human behavior. He criticizes the concentration of multistory dwellings near the Ribeira Palace, the center of government and commerce; the irrational behavior of evacuees, putting treasure above survival; and (implicitly) journalists' misplaced emphasis on misfortunes affecting the rich and powerful. And for Rousseau—who, in contrast to the celebrated Voltaire, toiled in obscurity and poverty—it is not surprising to see themes of class pervade each of his insights. Rousseau's argument moves the center of inquiry from the *physical hazard* to the *social risk*. Understanding physical hazard is the focus of Voltaire's nature-based argument; it suggests an alliance with the natural sciences—seismology, climatology, volcanology, and the like. Understanding social risk similarly relies on the natural sciences, but as we will see, it also requires significant investments in social science—psychology, geography, political science, economics—as well as a healthy dose of philosophy and ethics.

The story of Lisbon suggests a progression from theodicy to natural science and later to social science. The city's destruction roused many citizens from a complacency that had allowed them to grow too comfortable with aristocracy and vague notions of fate. In the aftermath, citizens demanded more of government and began seeing themselves as agents of change in their environment. In response, the Marquis de Pombal, a leading Portuguese statesman of the time, immersed himself in the practical details of reconstruction and launched one of the first scientific inquiries into the mechanics of earthquakes. Zoning rules were imposed, as were Europe's first seismic building codes. Walk the streets of Lisbon's elegant Pombaline Downtown and you will see what are considered the world's first earthquake-resistant buildings. The Marquis de Pombal is today regarded as a forerunner of modern seismology. Contrast the marquis with Gabriel Malagrida, the Jesuit priest, whose practice of blaming Lisbon's misfortunes on urban depravity so exasperated city officials that he was ejected from Portugal's royal court and banished from the capital. He would later fall into political scandal and be executed on charges of heresy.

Without doubt, the trend in American and international research leans strongly toward the social mechanism. Assessments of today's disaster risks have, according to sociologist Robert Bolin, correctly "shift the analysis of disasters away from the physical hazard agent and a temporally limited view of disasters as 'unique' events separate from the ongoing social order."[17]

In terms of climate change impacts, that means every risk a community faces includes both a set of *geophysical* variables as well as a set of *social* ones. The geophysical variables include things like rising seas, stronger storms, and parched, fire-prone forests. The social variables are more diverse and, in certain cases, even more influential than the geophysical factors. A major category here involves the built infrastructure. To assess a community's risk of flooding during a tropical storm, you need to know about the quality and location of housing, the adequacy of its roads and bridges, the fortification of its power plants. You will also be concerned about the availability of essential services—communications, medical treatment, insurance systems, and more. Finally, two key social contributors to a community's climate vulnerability are racism and poverty. We'll investigate these ailments later.

To sum up, we've looked at some of the ways people in other disciplines talk about resilience, noting that it always has something to do with "bouncing back," and, in its modern sense, often suggests change and improvement. Resilience is, thus, more than "toughness" or simple stability. When applied to human populations, resilience also has a prominent and inescapable social dimension, as do the disasters that foist misery on human populations. No natural disaster is purely "natural" as long as human communities are involved. And that's even before we get into the debate about what calamities can be attributed to human-made climate change. What's more, I think this is good news because while we have a hell of time controlling nature, we have at least some capacity to control ourselves.

The United Nations' Intergovernmental Panel on Climate Change describes climate resilience as "the ability to cope with a climate disturbance and recover in a way that preserves one's essential character, while at the same time exercising the capacity for adaptation, learning, and growth."[18] See the part about coping and recovering while preserving "one's essential character"? That's the bounce-back. Now look at "adaptation, learning, and growth." That's about change and improvement. Bounce back—but *better*.

As we saw in the previous chapter humans have been "bouncing back better" in dynamic environments for millennia. For the most part, they have all used the same "resilience recipe," which is made up of three strategies. They *resist*. They *adjust*. Or they *retreat*.

Imagine living in a coastal area that is less than thirty feet above sea level—the kind of place that ten percent of the world's population calls home. There are great reasons to live there—good fishing, easy access to trade, aesthetic, and cultural gifts. But it is risky too, with storms, churning coastlines, and now sea level rise.

Coastal communities can *resist* erosion and flooding with some form of engineering like levees, seawalls, jetties, or other infrastructure meant to keep the shoreline in place. So called "green" infrastructure like barrier islands, barrier reefs, and surge-slowing marshes also helps resist destructive forces to a surprising degree. Communities *adjust*, too. They elevate bridges and roads, install pumping systems, and make use of landscaping designed to retain and absorb stormwater. They impose special building codes, institute warning and evacuation protocols, and design insurance systems. *Retreat* means abandoning development that cannot reasonably be protected in another way. But it can also mean restricting new development in places where flood risks are already too high. These strategies sometimes overlap or are combined. Remember, though, that resistance, adjustment, and retreat, don't just happen. They are, ideally, planned in advance—often with community leaders working together toward a common goal. None of this is particularly new, although the climate crisis puts an obvious spin on things. Human beings have been resisting, adjusting, and retreating in response to their environments for millennia. And since the days of the Lisbon earthquake, we have done so with ever more awareness of the physical and cultural dynamics involved.

If you think the resilience recipe sounds simple, it's about to get more complicated. That is because people often disagree about how much of which ingredient—resistance, adjustment, or retreat—to put into the pot. Sometimes they disagree about whether some communities need stronger resilience at all. There are many reasons for such disagreements, but the main ones involve scientific uncertainty, bad economic incentives, and conflicting social values. Once you can recognize these problems in the climate context, you'll find you can't help seeing them again and again.

Scientific uncertainty is perhaps *the* defining feature of environmental policy. It's certainly true of climate action. We never know as much as we'd like before making choices that could remold the earth's surface in the next hundred years. According to surveys by U.S. climate officials, much of the resistance to building climate resilience now comes from the "the inability to predict future climate parameters with complete accuracy."[19] Will higher temperatures accelerate the melting of the Antarctic and Greenland ice sheets so as to add another three feet to end-of-century sea level predictions? Dunno. Climatologists say we need more science.[20] Will higher temperatures wreck the Gulf Stream and catapult northern Europe into a Winterfellian deep freeze? Again, not sure. Something something *science*.

Humans, by the way, have been fretting forever about acting without proper information. More than 2,500 years ago, the Persian prophet Zarathustra is thought to have advised, "In doubt whether an action be good or bad, abstain from it."[21] (Voltaire agreed and popularized the maxim in Europe.) Today in the context of environmental protection this idea has been kneaded and massaged into a doctrine called the "precautionary principle." A version of it appears in Article 3 of the United Nations Framework Convention on Climate Change, the 1993 agreement which the United States and 195 other nations have joined. It directs the parties to "take precautionary measures to . . . minimize the causes of climate change and mitigate its adverse effects." Where threats are "serious" or "irreversible," a "lack of full scientific certainty should not be used as a reason for postponing such measures."[22] Thus speaks Zarathustra—through the larynx of late twentieth-century diplomats.

The challenge is that the hardest questions are not about whether to act, but how far to go. Imagine you are a planner at Consolidated Edison, the electricity provider for New York City. You are charged with protecting the city's power plants from debilitating floods through 2070. During that time, the U.S. Department of Energy says your city can expect anywhere from one to four feet of sea level rise. With one foot of sea level rise, only one power plant is at risk. With four feet, nine plants could be swamped. Knowing your utility will want to pass its costs on to customers, how many of those plants should Con Edison fortify or move? When designing new plants, what new conditions should your engineers plan for? And by the way, is this even *your* decision to make? It sounds like the kind of exam question I use to needle my students, but it's not. It's real. In chapter 7, we'll consider one answer.

Another type of uncertainty involves scientists' ability to design resilience strategies that will work as advertised—or at least not make things worse by hurting the environment or pulling resources away from better ideas. If you could, would you move giant yuccas from the Mojave Desert to Utah where they might have a better chance of survival but could muck with the landscape? Would you go all in on a plan to resuscitate miles of dying corals with genetically altered algae? Before answering, read chapters 10 and 11.

Even when we are certain about what actions we want, we can still run into trouble when people get rewarded for doing the wrong things. This very common problem is what economists call "misaligned incentives." Say you want to make it easier in the future for victims of wildfire to rebuild their homes, so you join a campaign promoting state-subsidized fire insurance. That insurance will bring peace of mind to many, but it will also encourage people to move to or to continue to live in dangerous places like converted forest land. Misaligned incentives often arise from a "mismatch" in political jurisdictions or geographic scales. In chapter 8 we'll learn that a lot of fire-prone counties in the western United States don't do as much as they should to restrict residential development near forests. That's because expanding development increases the tax base, which increases local revenue. And the agencies most responsible for paying to extinguish those fires and to help displaced residents afterwards are agencies run by the *federal* government.[23] Fixing that problem would involve realigning incentives by matching governmental authorities and responsibilities in a better way.

Finally, people just care about different things to different degrees. At some basic level, we all want our communities to thrive and our planet to be livable. But we disagree about what that means. We have a plurality of values. One of the places this often comes up is in the contest between short-term and long-term goals. In Louisiana, we are losing our coast in part because the oil and gas industry has chewed through our wetlands like a hyperactive Pac-Man. But many are afraid of losing jobs if we restrict that exploration. According to current trends, my congressman is literally going to see the majority of his district fall into the Gulf of Mexico. But he is unwilling, out of concern for job loss, to do what it takes to slow the destruction.

That shorter-term view is easy to criticize. But look at me. I support a coastal resilience strategy in Louisiana that will cost upward of $100 billion over fifty years in hopes that the most populous coastal regions of the state will have another hundred years of flourishing. Don't ask me what happens after that because no one knows. I think it's worth it. Am I wrong not to care about the even longer term? You can decide when you get to chapter 6.

Scientific uncertainty, mismatched incentives, conflicting values: how can we tackle so many challenges at once? It would be easier to fold an octopus into a box! Maybe. But I'm throwing in my lot with the army of environmentalists, advocates, and community leaders, who say that societies can learn to adapt and improve, as even the crusty Jean-Jacques Rousseau seemed to believe (on his good days). If that's right, we'll need a few principles to point us in the right direction. And maybe a clever octopus to keep us inspired.

Captivated by my conversations with Sealey, I binged a few other documentaries on sea life, my eyes fixed limpet-like on the flatscreen TV. In a streamed episode of *Blue Planet II,* the popular BBC production narrated by David Attenborough, I watch an octopus caught outside its den about to be attacked by a pyjama shark. The octopus suckers up a jumble of shells and folds itself inside of them, looking something like a crumpled soccer ball. The shark circles once and butts the soccer ball with its snout. The octopus contracts. The ball's panels inch closer together. It shimmies a little to the left. The shark stares, tilts its head, and pokes again. This time the octopus jets off with a puff of sand. "In a forest full of hungry mouths," Attenborough intones, "superior wits allow this octopus to stay alive."[24] Resist. Adjust. Retreat.

For human beings, resilience need not mean suckering up clam shells (although to armor shoreline, it might be worth a try). The octopus's trademark characteristics, however, can help us remember some important lessons as we prepare for climate change. I'm thinking, specifically, of three traits. First, the octopus is a case study in situational awareness and flexible thinking. Most species of octopus live only a year. In that time they are continually surveying the territory and experimenting with whatever resources lie in arm's reach. That is the lesson of the lobster hunt in *My Octopus Teacher.* If you need calories and your familiar fuel is unavailable, find something else. If your first attempts to capture prey fail, learn from your failure then try something new. "Octopuses," writes Peter Godfrey-Smith, a philosopher and amateur diver, "have an opportunistic, exploratory style of interaction with the world. They are curious, embracing novelty, protean in behavior as well as in body."[25]

We humans might think of ourselves as protean, but we easily fall into ruts. That's especially true of our laws and governing structures. In fact, laws and governing structures are *supposed* to push us into ruts. They are meant to stabilize collective living so we know what is expected of us and what we can expect from others. We don't like it when people in government tell us to rethink our diets, our modes of transportation, or the places we live. As for exploration and novelty, political leaders would sooner wrestle a pyjama shark then sell *that* to constituents.

We need to imagine ways to govern that provide sufficient stability while remaining flexible enough to adapt to new information and new circumstances. Integrating curiosity and novelty into climate policy is a major theme of this book. If that kind of flexibility makes you nervous, think of it as what activist Greta Thunberg has called "cathedral thinking." That is, "we must lay the foundation while we may not know exactly how to build the ceiling."[26]

Being flexible also means understanding that what presents as a climate problem may, in fact, have multiple layers. Take coral bleaching, a global crisis we'll explore in chapter 11. It's true that without human carbon emissions, corals would not be dying at such alarming rates. But other factors are contributing too, among them over-fishing and land-based pollution. We can't immediately cool the ocean. But enforcing fishing laws or limiting agricultural runoff might lighten the load enough to let some reefs cope and heal. Human communities have social stressors like poverty, racism, and political disenfranchisement that can undercut *their* ability to cope and heal. Flexible systems allow people to address problems at different layers in order to ensure the best outcomes.

Second, the octopus reminds us to remain *forward-looking* when we plan. I told you about one octopus's shell-encrusted soccer ball. Peter Godfrey-Smith reports that in 2009 researchers in Indonesia were surprised to find octopuses lugging pairs of half coconut shells—one nested in the other—as a portable safe house. When a threat arose, the octopus would assemble the halves into a sphere and hide inside. Lots of animals build shelters from found objects or use sticks or rocks to collect food. But to devise and travel with a multipiece, collapsible vault is pretty unusual. As Godfrey-Smith writes, "The coconut-house behavior illustrates what I see as the distinctive feature of octopus intelligence; it makes clear the *way* they have become smart animals."[27] The octopus is not just experimenting with novelty—it is thinking ahead. The animal has to anticipate future conditions to have the motivation to shuttle those awkward coconut halves along the seabed.

Human beings make lots of plans for the future. Our bulging prefrontal cortex is custom-built for that. But where climate breakdown is concerned, we tend to plan for the weather impacts we have seen years before, rather than the new, more extreme ones, that our cortexes tell us are coming. There are explanations for that. As we saw earlier, it's hard to plan when there is so much scientific uncertainty in our forecasts. Reasoning, too, is partly an emotional process; and letting go of old assumptions and values challenges our desire for identity and belonging. Fortunately, there are ways to surmount both of these challenges, and in upcoming chapters we'll see how.

Another part of looking forward is making sure that while you are pursuing a fix, you aren't simultaneously making something else worse. Resilience gurus call this "maladaptation." Consider the city of Miami Beach. At the beginning of this book I mentioned that the municipality was addressing the problem of "sunny day" floods by installing an advanced pumping system. Because the sea is rising, stormwater can no longer be drained away by the force of gravity. The trouble is this pumping system will use tremendous amounts of electricity; and if that electricity is generated by burning natural gas or another fossil fuel, Miami Beach's climate resilience efforts will further contribute to global warming. Because so many resilience technologies rely on electricity, from air conditioning to desalination plants, plugging more carbon-free generation into the power grid is a top priority.

Sometimes one kind of resilience impinges on another kind. Later on we'll see that solar thermal plants, which require large amounts of water in their cooling process, do a good job of reducing carbon but also contribute to problems of climate-driven water scarcity. My home in New Orleans is protected from climate-driven storm surge by a sprawling system of levees that has partially contributed (in addition to oil development and other causes) to the disintegration of the vast wetlands that also buffer us from storms. Sometimes these trade-offs may make sense, but the key is they must be considered as part of the original plan. Looking ahead with blinders on is not allowed.

Were an octopus to extend its rubbery limb into, say, the hollows of a drain pipe belonging to a condominium's parking structure, science tells us that such a pedestrian act—the mere groping for food—would give rise to a "mixture of localized and top-down" cognitive control that is exceedingly rare in the rest of the animal kingdom.[28] Put another way: the creature's brain and its arm would both be thinking at the same time. I don't mean this in a figurative sense, as when I swing a backhand on the tennis court with the impression that my right arm "just knows what to do." I mean it literally.

Octopuses have little knots of neurons distributed among their eight limbs. These knots, called "ganglia," are essentially small brains. Thus equipped, each arm has not only its own sense of touch, but the capacity to sense light, to smell, and to taste. In the laboratory, an arm that has been surgically removed has been observed performing "basic motions, like reaching and grasping" all on its own. Octopuses also have a larger central brain, housed in the mantle that also directs movement and keeps track of things, though scientists are still not sure how these brains interact with one other—or even if interacting is the best way to describe what they're doing. This contrasts with the standard view of cognition, which sees an animal's knowledge as belonging to *either* a CEO-like brain or to various body parts.

Chalk it up to my being a law professor, but upon learning of the overlapping relationship between octopus mind and body, my thoughts immediately leaped to the U.S. federal system of government. Our government operates on many levels (federal, state, local, and tribal) and across many sectors (transportation, agriculture, environmental protection, and so on). Climate resilience requires this kind of sharing relationship among our governing "ganglia." In other words, operations and responsibilities must be *fit to scale*. We will see that many climate resilience efforts emphasize decision making and implementation at the local level. That makes sense because climate impacts vary with geography and regional culture. Because of the urgency of the challenge and the lack of proven methods, experimentation—much of it occurring at the local level—will be critically important. Decentralization can also leverage local knowledge and take better account of community preferences. We must also, however, be on the lookout for circumstances where the current fit doesn't work. Think of the fire-prone counties I talked about that invite revenue-generating development and leave it to the federal firefighters to protect residents from wildland blazes.

Our success in building climate resilience will depend on systems that are *flexible*, *forward-looking*, and *fit to scale*. To that I will add one more trait, to which the next chapter is entirely devoted. To work as advertised, climate-resilience efforts must also be *fair*.

chapter 4
Climate and Caste

Some days after Hurricane Ida, in the steam-roasted Louisiana parish of St. James, I joined a group of volunteers in the morning at a local park to help distribute ice. The place had been battered. Trees and utility poles were down everywhere. Decapitated homes lay exposed to the elements. Sheets of metal swung from the skeletons of fast-food signs. For nearly a week, people had been without electricity. A local company had donated three pallets of crushed ice, packed in seven-pound plastic bags.

The event had been orchestrated by Sharon Lavigne, the founder of a grass-roots environmental justice organization called RISE St. James. Lavigne, a retired special-education teacher, has lived in St. James Parish her whole life—more than seventy years. Her home is part of a ribbon of settlements, mostly Black, Latino, and poor, that curve southeast along the Mississippi River from Baton Rouge to New Orleans. With more than two hundred refineries and chemical plants, this eighty-five-mile stretch is home to a quarter of all petrochemical production in the country.

As I sling two bags of ice onto a neighbor's flatbed, I notice Lavigne stepping out of her muddy SUV a few yards away. Even with Covid masks, we recognize each other. She has spoken to my environmental law students many times as part of a bicycle tour my class takes each year through the chemical corridor. She's wearing a pink-and-white plaid shirt and a black visor bearing the "RISE St. James" logo, which features both a Latin cross and a Black Power fist. Lavigne greets me with smiling eyes and a hug. I ask how she made out during the storm.

"It was the worst I've ever seen," she says.

On August 29, 2021, the Category 4 storm had pummeled the coast with peak gusts of 172 miles per hour and up to nine feet of storm surge, flooding communities and destroying homes. The hurricane displaced thousands of people and left more than a million without power—all as a surge in coronavirus cases overwhelmed hospitals across the state. In St. James, about fifty miles inland, the storm's rotating bands of wind and rain hovered for several hours.

The walls of Lavigne's house, which are made of brick, withstood the wind. But large sections of the metal roof were torn away. "You could hear it when the wind pulled it off," she tells me. "I stood by my window, and I looked. I said, 'Dear Lord, don't let it bust my window.' And He didn't. But you should have heard the water coming through the roof. It was pouring through like a faucet." Lavigne emerged in the morning to find most of her ceilings caved in and the contents of her home soaked. "It's all messed up," she says, "but I'm good. I'm here."

Lavigne checks in with the other volunteers, who come mostly from surrounding neighborhoods. Like her, they have put their own storm recovery on hold so they can help those who are in even worse shape. The pallets of ice had been delivered right on time this morning. In the afternoon, a U-Haul truck from Mississippi was expected to arrive with cases of canned food and breakfast cereal, all donated from private food banks. (Government-sponsored support, whether from federal or state agencies, has been slow and unreliable.) A volunteer crew of high school and college students has started their first shift clearing storm debris from the yards of elderly neighbors.

Lavigne is happy about this but also worried about the continuing release of industrial gases in the area. These kinds of discharges typically follow storms like Ida. Some are caused by fires or explosions at damaged facilities. Some are intentionally triggered. This can happen when a plant's electricity source fails and operators are forced to purge pressurized chemicals to avoid such explosions. In the week following Ida's landfall, more than 170 reports of chemical discharges were filed with the EPA, many of them involving airborne releases in Louisiana. No one really knows how much poison was pumped into the Louisiana sky that week, which is thought to have included ammonia, hydrogen sulfide, and gasoline vapor. When the power was knocked out, the state's air monitoring equipment quit.

Generations ago, locals referred to the region as "Plantation Country" because of the acres of sugarcane, mansions, and forced-labor camps that dominated the area.[1] Today, it's known as "Cancer Alley" in recognition of the decades of industrial pollution that have driven local cancer risk through the roof, the

4.1 Sharon Lavigne, founder of RISE St. James.
(Courtesy of the Goldman Environmental Prize)

highest in the country, according to the U.S. Environmental Protection Agency (EPA).[2] About fifty miles from the coast, Cancer Alley is also notoriously prone to floods and storms, made more intense and unpredictable by an unraveling climate.

Lavigne tells me that there are two flares coming from a plant that makes polystyrene, the stuff Styrofoam is made of. She wants to take me to see it and make a video to post on her Facebook feed. In minutes we're racing up Highway 18 along the river toward the plant. It's this enthusiasm, backed by religious faith, that explains Lavigne's success as an environmental advocate. The matriarch of her family, with six children and twelve grandchildren, Lavigne also knows a little something about organizing and motivating people. When a giant plastics manufacturing company wanted to construct a plant near her home, she founded RISE St. James and led a campaign to stop it. For that effort and her continuing work on climate action and racial equity, Lavigne was awarded the Goldman Environmental Prize, known internationally as the "Green Nobel" (figure 4.1).

We arrive at the plant. Flakes of gravel pop under the tires as Lavigne swings her SUV into the parking lot.

"The air smells horrible," she says.

We get out, and Lavigne hands me her cell phone. I'm going to be the camera man, she tells me. I frame an image of her standing a few feet from a mass of tanks and metal towers surrounded by a chain-link fence. In the distance, two amber flares dance above her head. I signal that I'm recording.

"Good morning," she says, polite but serious. "I'm Sharon Lavigne from St. James, Louisiana. I'm calling to report flares from two sites. These flares have been going on since after the hurricane. I'm calling on EPA to investigate this facility. It's called American Styrene. Today is Saturday, September 11, 2021. Thank you."

You can draw a straight line from storm damage in St. James Parish and emergency discharges from Styrofoam plants to climate breakdown. A longer chain stretches centuries back, across the torturous Middle Passage, cradle-to-grave bondage, abandoned Reconstruction, Jim Crow segregation, officially sanctioned disenfranchisement, and, finally, the many sorrows faced by Lavigne and her neighbors today, among them cyclical poverty, industrial pollution, bad health, shoddy housing, and anemic emergency-response services. As researchers at the University of London aptly noted in their study of this river corridor, "environmental degradation and cancer risk manifest as the by-products of colonialism and slavery."[3]

Climate change and subordination each add a viscid layer of risk to the community-hazard portfolio we saw in chapter 3, whether by amplifying an external threat or weakening people's ability to cope with it. In the words of United Nations Secretary-General António Guterres, "Climate change is happening now and to all of us. . . . And, as is always the case, the poor and vulnerable are the first to suffer and the worst hit."[4] That the "poor and vulnerable"—whose activities consume less fossil fuel per capita than the rest of society—have contributed least to the crisis now threatening their lives is at once unjust and completely predictable.

Studies in the United States show that on every point of the disaster timeline—from preparation to response to recovery—socially imposed vulnerability loads the dice.[5] Demographic characteristics that have been linked to disaster vulnerability in the United States and in a variety of other countries include race, class, sex, sexual orientation or gender identity, age, and disability.[6]

For instance, poor people and those with racially restrictive housing choices are more likely to live in flood plains or in housing that lacks proper insulation, air conditioning, or other features. Many studies show that the Federal Emergency Management Agency (FEMA), charged with helping Americans recover from extreme events, often helps white disaster victims more than people of color, even when the damage is similar.[7] That's especially bad news given that in the United States, poor Black and Latino counties are among those researchers say will suffer the most on account of climate change.[8] Perhaps for this reason, people of color in the United States are far less likely to be politically polarized on climate change than whites.[9]

In addition to race, language barriers make it harder for some immigrants to get early information about weather-related threats and to later navigate compensation and recovery processes. Fear of immigration officials also discourages some families from accessing evacuation and other emergency services they are entitled to.

Research shows that women are often more likely to lose their jobs after disasters involving fire and flood, while prospects for men increase as the demand for repair and construction soars. When children's schools are disrupted, it is women who primarily take up the job of home schooling and childcare.[10]

Gay and transgender people say that they, too, bear special burdens as climate impacts increase. Aletta Brady, a climate advocate and self-described "queer writer," notes that because gay and transgender people tend to be concentrated in coastal cities like San Francisco, New Orleans, and Miami, they are at disproportionate risk from climate-inflected floods and storms. On top of that, social prejudice adds a second layer of economic and emotional stress, particularly for teens and young adults. "Many young queer and trans people," she and her co-authors write in *Grist*, "do not have the resources or ability to flee verbal, emotional, and physical violence in their hometowns. Instead, around the world, they are often abandoned by family and forced onto the streets with little to no support to survive." Brady emphasizes that 40 percent of youths in the U.S. homeless population identify as LGBTQ, while making up just 7 percent of the general population.[11]

Older adults and people with chronic illnesses or disabilities often have a harder time safely evacuating from fires, storms, and other threats. They are also more vulnerable to episodes of heat stress and exposure to smoke from wildfires.

In these examples, one might say the less physically robust are more "naturally" susceptible to climate risk. But in the spirit of Rousseau's framing of the Lisbon quake, I want to challenge that. On our own, each one of us is supremely vulnerable to the ferocity of nature, "red in tooth and claw," as the British poet Alfred Tennyson wrote.[12] How well we survive that situation has a lot to do with the support and protection society offers. As much as any innate ability, it is our shared support systems that separate the weak from the strong. And in the United States, many of those systems are shaped by the underpinnings of social caste.

The term *caste*, of course, evokes the ancient hierarchical structure that still presides over many parts of the Indian subcontinent. It is, perhaps, the world's oldest existing mode of social stratification. Hindu texts speak of four tiers, or *varnas*, used to denote an elaborate social pyramid. On top are priests and teachers (Brahmans), followed by rulers and soldiers (Kshatriyas), then merchants (Vaisyas), and laborers (Sudras). Below them are Dalits, a class of more than two hundred million Hindus who technically live outside the caste system. They are meant to take on the dirtiest and most unhealthful jobs—sorting garbage, clearing sewers, tanning leather, and the like. India's constitution bars discrimination based on caste—and some Dalits do manage to climb the nation's professional and economic ladders. But in hundreds of thousands of villages across the subcontinent, caste defines the social pecking order. In those communities, Dalits are often expected to use separate drinking fountains, dine at separate tables, and dwell in poorly serviced, deeply segregated neighborhoods. When they die, many are buried in separate cemeteries.

I've lived in India on two occasions in my life—once as a recent law school graduate with long hair and backpack and again, years later, as a Fulbright researcher. I consider it one of the most fascinating and sublime places on the Earth. But I never passed a day in the country's streets or valleys on which I did not grieve over the pervasive power of those cultural shackles.

It always reminded of me of America's repulsive race-based hierarchies. My impression was hardly new. Intellectuals ranging from the anthropologist Gerald Berreman to the civil rights lawyer Michelle Alexander have drawn parallels between India's caste system and the Jim Crow South.[13] The journalist Isabel Wilkerson recently extended that reasoning into the present day and even beyond race. In her book *Caste: The Origins of Our Discontents*, she argues that the

American caste system never disappeared and that its reach extends across the entire landscape of U.S. policy and culture.

Wilkerson defines casteism as "any action or institution that . . . seeks to keep someone in their place by elevating or denigrating that person on the basis of their perceived category." In the United States, the most destructive and tragic rankings involve race and ethnicity. But American casteism, Wilkerson argues, also draws on the categories of gender, LGBTQ status, age, disability, and other personal characteristics. Not everyone would define caste so broadly, but the idea is helpful, I think, in understanding climate risk. Historically, caste hierarchy has been so tightly integrated into our legal frameworks and cultural sensibilities that many people on the higher rungs no longer see it. Or they have stopped paying attention. Yet casteism persists, fueled by what Wilkerson describes as the "automatic, unconscious, reflexive response to expectations from a thousand imaging inputs."[14]

When the chemical plants in Cancer Alley forced open their vents to blast hydrogen sulfide into the air, it didn't occur to anyone in charge to warn fence-line residents that this toxic brew might send them to the hospital. In "Plantation Country," the residents, poor and Black, have learned never to expect such things. They don't even expect that federal recovery staff, who are often transported from neighboring states, will know anything about the local pollution they fight, the community's preexisting medical conditions, or the generations of poverty and racism that underlie these ills. Judging from my many conversations with storm victims after Hurricane Ida, that sort of institutional ignorance is still pretty common. For many Americans (including those with the best intentions), it's too easy to view communities like St. James as unremarkable, hard-luck cases—examples of the way things "just are." The Indian jurist and activist B. M. Ambedkar, himself a Dalit, observed that "political tyranny is nothing compared to social tyranny."[15] The leviathan is everywhere. "No one," writes Wilkerson, "escapes its tentacles."[16]

While these bonds affect us each in sometimes dramatically different ways, each of us has the moral duty to do what we can with the capabilities that we have to help subdue the monster. Lavigne, who came of age as a Black girl in segregated Louisiana, now leverages her faith, her gift with language, and her standing in the community as a former teacher. My background, of course, is very different. I was a white boy growing up in suburban Las Vegas—a kid whose parents and grandparents had steady jobs working in the casinos. Yes, that was a challenging world to grow up in and, yes, like many of my peers, most adults I knew lacked college degrees. But I also enjoyed a certain level of social and economic privilege (thank goodness for unions) and caught a lot of lucky breaks. Today I try to leverage what I have—my curiosity, my platform as a professor, the persistence I learned from my family—to make the world a smidgeon better, which is to say, more just and, therefore, more resilient. In different ways, that's something we can all do.

———◆———

One evening in New Delhi, on the ride home from my office, I asked my auto-rickshaw driver about climate change. He was a chatty guy, bearded, with black spectacles and a blue turban. Like cabbies around the world, he had a lot of opinions. "Too many people!" he shouted, his voice competing with the vehicle's rattling frame and the bleats of oncoming horns. "Too much traffic!"

We swerved around a traffic circle. I gripped my seat. A copse of date palms swung by, then a billboard. "Enrich Delhi's Green Legacy," it said. I took the bait. "Hey, do you worry about global warming?" I shouted. We slowed to a stop behind a row of cars and two-wheelers waiting for the light. He cut the motor. A boy pranced into the traffic and began turning cartwheels for remuneration.

"Yes, I know about that," the driver said. "Too much warming. Too much heat."

"But do you worry about it?"

"Me? No." He fired the engine and frowned slightly. "You know, India has too much noise!" he shouted. "And too many dogs. Too much *everything*!"

It's true that India has a lot going on. That's one of the reasons that, a few years back, my family and I had decided to move there for a semester, where I studied government resilience projects and gave lectures on climate policy. With 4,000 miles (7,000 km) of coastline, Himalayan glaciers, and nearly 270,000 square miles (70 million ha) of forests, India is especially vulnerable to climbing temperatures, higher seas, and stronger downpours. And with 1.3 billion people, an economy worth over a trillion dollars, and, yes, at least thirty-five million stray dogs, there is an awful lot at stake.[17]

During that time, I traveled widely and visited many unusual sites. But one place—the Hazira Industrial Area in the western state of Gujarat—struck me as very familiar. It looked like Cancer Alley. This stretch of refineries, factories, and villages, referred to by local boosters as "The Golden Corridor," follows National Highway 6 as it winds westward along the Tapti River from the city of Surat to a busy trading port on the Arabian Sea. Half a century ago, except for a few settlements, the place was all forest and swamp. Then in the 1980s, the Gujarat government opened the region to industrial development. When India liberalized its trade policies in 1991, foreign investment from petroleum companies began flowing in, and waves of migrant labor followed. About fifteen miles upriver is the city of Surat, with a population of over six million. Situated along a crescent-shaped bend on the commercially important Tapti River, Surat resembles an overbuilt New Orleans when viewed on a map. And like the Mississippi River, this "strong brown god" is prone to some mighty flooding.[18]

During my first visit there, I was in the front passenger seat of white sedan, accompanied by two planning experts, Anup Karanth and Mehul Patel. Cruising west on Highway 6, we passed scores of industrial facilities, each animated by chugging pumps, flaring plumes, and brown clouds of toxic gas. I could see names like Shell, Essar, and Gujarat State Petroleum. We pulled over on the side of the road to get a closer look. On one side of the highway was a line of hulking structures, all surrounded by brick walls and concertina wire. On the other side was a shantytown that seemed to tumble out of the forest behind, its architectural elements dominated by blue tarping, corrugated metal, and bicycle tires. I watched two women in saris heaving pickaxes into a muddy trench. A little egret took flight, its golden slippers reflecting the morning sun. "To be very honest," said Karanth, "it's the craziest place on earth."

Crazy—and wildly exploited. The refining and manufacturing activities, which drain metals like nickel, copper, and cobalt into the surface waters, have gravely damaged the estuaries upon which fisherfolk rely. Leaks and fires pollute the air and threaten local safety. In 2020, a triple explosion at a Hazira gas-processing plant in the middle of the night sent workers and villagers scattering, though, thankfully, no one was killed.[19] Dredging activities related to the port, along with an expansive sand-mining industry, are eroding the shorelines and causing the whole region to sink, even as the sea continues to rise. According to Patel, when the region floods, as it does every few years, the shanties all get washed away. In 2006, the entire region was submerged for several days, he told me. Of course, even smaller floods will carry contaminated soils across the highway and onto the dirt floors of people's homes. The nearby estuaries, from which fish and shellfish are pulled, are choked with sediment and dangerous levels of nickel, cobalt, and chromium.[20]

In contrast to the residents of Cancer Alley, the people of the Hazira Industrial Area are mostly newcomers. Many have left failing farms, punished by drought, infestation, and other ills that scientists say are being made worse by global warming. Their lack of employment and community ties makes it much harder for them to find shelter in Surat's urban core. So they come to Hazira and settle on any puddle, footpath, or shoulder of a train track that hasn't already been claimed. Many hope to do well enough here to allow them to move upriver to Surat, where wages are higher. It's a bustling city, internationally known for textile manufacturing and diamond-polishing. Surat's shantytowns, built along the river's crescent bend, are moderately better than those in Hazira, with at least some access to toilets and tap water. But during the monsoon these settlements also flood, sometimes demolishing whole neighborhoods and leaving many families worse off than before they left the farm.

It's tempting to attribute this narrative to the vagaries of fate, the snakes and ladders of human endeavor. But, as in Louisiana's Cancer Alley, this winding narrative hangs on the lattice of social tyranny. It won't surprise you to learn that the people who come to the Golden Corridor are mostly poor and with little formal education. Many are also either members of religious minorities, Hindus of lower caste, or Dalits.

Seen from a global perspective, the reinforcing effects of climate and caste are almost uncanny. To begin with, the physical toll of climate disruption is already weighted against those with the fewest resources. That's because while the climate crisis threatens people everywhere, much of the worst damage will be in the tropics. Inhabitants of these regions, who account for more than 40 percent of the world's population, are mostly people of color and of lower income. You may already know that cyclones, monsoons, and the cycles of El Niño and La Niña have their strongest and most disruptive effects in tropical regions, which extend about 23 degrees northward and southward from the equator. (That includes much of South America, sub-Saharan Africa, the subcontinent, and Southeast Asia, as well as many of the Pacific islands.) There is persuasive evidence that global warming will intensify all of these events.[21] For geophysical reasons, sea level rise is more pronounced near the equator and poses an existential threat to many small islands in the Pacific. Farming in the tropics is also hard going, not only because of big storms and variable rainfall, but also because of poor soil, the abundance of pests, and too much disease. Global warming is making most of those things worse, too.

People in poorer countries, such as those in the tropics, are least able to cope with sudden disasters when they do occur because of substandard infrastructure, poor emergency response, and insufficient medical services. Indeed, deaths from natural disaster are generally linked to a nation's economic welfare and its degree of income inequality.[22]

Poorer nations also show greater sensitivity to climate change because of their economic reliance on agriculture and fragile ecosystems. In India and in sub-Saharan African nations, more than half the population is engaged in rural farming.[23] Compared to more affluent countries near the poles, a large share of the population is made up of infants and children, elderly people, and the disabled. That means more work for women and girls, who often care for those in need. Women and girls, who in many cultures also grow the food and collect the water, will also find these chores consuming more of their time, leaving less time for education and vocational training. By pushing girls out of school, climate change is expected to increase female illiteracy

Research from two climate scientists at Stanford University, Noah Diffenbaugh and Marshall Burke, suggests that global warming is already making poor tropical countries substantially poorer by pushing them deeper into oven temperatures. And what about those richer, wealthy countries near the North Pole? Evidence suggests that some of them are being made richer! (Longer growing seasons, less snow to plow, and other advantages.) For instance, Diffenbaugh and Burke calculate that India's economy has shrunk more than 30 percent on account of climate breakdown, while Norway's has grown 30 percent for the same reason. That story is playing out across the globe, from Sudan to Sweden, from Columbia to Canada. "We find very high likelihood," the scientists write, "that anthropogenic [human-caused] climate change has increased economic inequality between countries."[24] The situation looks even worse, when you consider that, per capita, residents of poor tropical countries have a disproportionately low responsibility for causing greenhouse-gas emissions in the first place.

Of course, climate change didn't create global inequality. The causes of that run deep, among them geography, disease, technology, culture, colonialism, slavery, and modern policies of extractive trade.[25] The last four items often or always depend on systems of hierarchy, or to use Alexander's framing, "elevating or denigrating" people "on the basis of their perceived category." That is to say, caste. Like it or not, a lot of human history is about some groups of people trampling on other groups of people and using narratives of innate superiority and pecking order to justify it.

To sum up, we've seen how climate and caste team up to tackle already disadvantaged groups, bruising them up and keeping them down. This happens in rich countries like the United States, poorer countries like India, and among players in the global community at large. What should we make of this? And what should we do?

To answer these questions, I want to draw from the work of two modern philosophers who have often helped my students and me grapple with issues of fairness. The first is Judith Shklar, a political theorist who wrote about democracy and nationalism, often informed by her own teenage experience as a Jewish refugee from wartime Europe. While other theorists of her generation spent their energy tracing the perfect silhouette of "justice," Shklar took up what she thought was the more efficacious task of spotlighting "injustice," then showed how to avoid it.

According to Shklar, you can take all of the bad things that happen in this world and throw them into two buckets, one labeled "Misfortune" and the other "Injustice." In a prestigious lecture delivered at Yale in 1988, Shklar asked listeners to imagine an earthquake. "If the dreadful event is caused by the external forces of nature, it is a misfortune and we must resign ourselves to our suffering," she said. "Should, however, some ill-intentioned agent, human or supernatural, have brought it about, then it is an injustice and we may express indignation and outrage."[26]

Earthquake? Ill-intentioned agent? Outrage? If this makes you think of Rousseau and the Lisbon quake from the last chapter, you are in good company, because Shklar was probably thinking about that too. Jean-Jacques was one of her favorite writers. And like Rousseau, she had high standards for her fellow humans. "That something is the work of nature or of an invisible social hand," would later write, "does not absolve us from the responsibility to repair the damage and to prevent its recurrence as much as possible. "It is not the origin of the injury, but the possibility of preventing and reducing costs," she continues, "that allows us to judge whether there was or was not unjustified passivity in the face of disaster."[27] An important aspect of Shklar's method is that one should assess the fairness of a situation by taking the perspective of the least advantaged members of society. One reason for that is because, in practice, that perspective is the one most often left out. In human dealings, it's just too easy to blame the victim.

So maybe the earthquake had a natural cause, but the city might have responded better. And even if Hurricane Ida was inevitable, the Styrofoam plant could have avoided venting toxic exhaust in Black neighborhoods by shutting down earlier. And maybe Hazira was always going to be a gooey marshland, but that doesn't mean you entice the nation's poorest and most dismissed groups to live there. Anyway, once they are there, society owes *some* duty to protect them. And because science tells us that global warming is, in fact, supercharging Gulf hurricanes and prompting more floods in South Asia, those who have consumed the most in the industrial era—thus producing more carbon emissions—owe an even greater duty "to prevent" harms from recurring "as much as possible."

While Sharon Lavigne and I were driving toward the American Styrene plant in St. James Parish, Lavigne told me a story about her brother Milton, who rode out Hurricane Ida in his ranch-style house. The house did okay, but Milton, who uses a wheelchair, was trapped in his home without power for nearly a day because a large tree had fallen across his driveway. Neither his wheelchair nor his SUV could get past it. A parish cleanup crew investigated, but the workers refused to touch the tree because it was on private property. (A liability issue, they said.) It took hours of negotiation and jawboning, with neighbors and family all joining in, before the crew finally agreed to remove it.

I think about this story now because it raises a question Shklar's analysis does not completely answer. We might agree that government owes some duty to protect people, to "repair the damage," as she says, or even to help them "build back better." But how much support are we talking about? "As much as possible" is a good line, but what does that actually mean? In my gut, I believe the parish had a duty to free Milton from his house—to restore his liberty, if you will—by finding a way to swiftly dispose of that tree. I don't say the parish had to do that for *everyone* whose driveway was blocked. But Milton's needs were different.

In addition, I find it relevant that Milton is an older Black man who grew up in the Jim Crow South. These facts may or may not make his claim to a clear driveway more compelling. But they present a context that I don't think society should ignore. In the United States, race, gender, age, health, and mobility are "perceived categories" (to use Wilkerson's phrase) that have long been deployed to keep people "in their place." It turns out, Milton's well-being and his ability to build back better are about more than one lousy tree in the driveway.

To better explain this intuitive response, I need more precise language and probably a clearer rationale. So let me turn to Amartya Sen. He is an Indian philosopher and economist, famous for his research on social inequality, who has spent most of his career teaching in England and the United States. His work, for which he received a Nobel Prize in 1998, helped transform the way the United Nations measures poverty and inequality around the world. To measure a population's well-being, Sen focuses on what he calls "basic capability."[28] Rather than look just at material wealth or laws promising equal treatment, Sen asks what people are capable of doing or becoming *in real life*. That means taking into consideration personal characteristics like sex and body size; environmental conditions like climate and disease; and systemic prejudices like racism, sexism, and casteism. The goal is to ensure that every person, regardless of situation, has the ability to pursue "a life worth living which one has reason to value."[29]

We might disagree about what exactly that goal requires. But at the very least Sen thinks all persons should have the means "to avoid . . . starvation, undernourishment, escapable morbidity and premature mortality," to become "literate and numerate," and to enjoy "political participation and uncensored speech."[30] You might recognize this concept in current discussions about social justice and race relations in the United States, where progressives argue for *equity* in addition to *equality*. In this usage, *equality* means treating everyone the same way regardless of their situation, while *equity* means giving each person what they need to have a fair shot at success. As one advocacy group puts it, "Equality is giving everyone the same pair of shoes. Equity is giving everyone a pair of shoes that fits."[31]

What do you get when you mix Shklar's emphasis on government responsibility with Sen's vision of equity and well-being? A recipe for climate resilience with justice at the core. Shklar shows us that building resilience ("to repair the damage and to prevent its recurrence") is not just a nice thing for government to do, but an obligatory step toward *justice*. Sen reminds us that because we all live in different situations—bounded by personal characteristics, environments, and bigotry—we will sometimes need different kinds of support, even as we pursue the same dream: a life worth living, free of escapable injury and premature death.

I want to make two points that follow from this discussion very explicit. First, a climate-and-caste approach to resilience requires meaningful participation from historically disadvantaged groups. Shklar implies this in her injunction to see the world from the perspective of the powerless. Sen says it outright. He knows that the only way to be sure that those on lower social rungs will be accounted for is to make sure they have real-world political influence. To make the point, he often notes that there has never been a major famine in a functioning democracy, no matter how poor As he explains, "Authoritarian rulers, who are themselves rarely affected by famines . . . tend to lack the incentive to take timely measures. Democratic governments, in contrast, have to win elections and face public criticism, and have strong incentives to undertake measures to avert famines and other such catastrophes."[32]

The second point is that resilience requires more than offsetting caste-based oppression on the back end by considering a population's unique disadvantages. It requires eradication of this social tyranny on the front end, that is—to use a phrase popularized by Ambedkar and other pro-Dalit activists—the very "annihilation of caste."[33] One reason is that raising the alarm of burning forests and sinking islands requires the unfettered advocacy of the people most at risk. That advocacy, in turn, requires unfettered access to financial means, knowledge of facts and processes, and meaningful access to the polls. Eradicating systemic racism and all the rest is a heavy lift, I grant (and well beyond the scope of this book). But as we contemplate the road toward climate resilience, it is important that we see our journey as requiring progress on the social-justice front.

What this means for Sharon Lavigne's brother, Milton, and the other residents of St. James is that they are entitled to the protection and support of their government agencies with due consideration of their specific capabilities and needs. Critically, the dynamics of caste-based oppression cannot be ignored. Caste-based oppression, which continues to stunt the resilience and growth of our communities, must be called out, confronted, and dismantled.

I'm not saying that everyone is entitled to the same insurance policy, home generator, or seawall. But in the era of climate breakdown, everyone is entitled to have access to what is needed to avoid escapable injury and death, to pursue a livelihood in a rapidly changing economy, and take part in decisions concerning public health, resource conservation, managed retreat, and more. Everyone gets a fair shot at this, regardless of the special challenges imposed by race, class, disability, or geography.

To be sure, there is some progress being made. For instance, under the Biden administration, FEMA updated its individual assistance programs to better meet the needs of the poor, people of color, and other "underserved communities." The new policies make it easier for victims of disaster to hire mold-removal specialists, to install wheelchair ramps and grab bars, and to verify the ownership of a damaged home where records of clear title are lacking, a common problem in some Black and Latino communities.[34]

More dramatic actions are packed into the two new laws funding infrastructure and climate initiatives, which I mentioned in chapter 1. The Infrastructure Investment and Jobs Act will invest tens of billions of dollars to repair crumbling wastewater, drinking water, and stormwater systems. That could go a long way toward preparing underserved communities for climate disruption. Historically, the problem with investment programs like these is that they rely on a competitive grant process, which tends to favor communities that have the staff and resources to whip up exciting, well-researched applications. Poorer communities get squeezed out. In response, the Biden administration has launched a special initiative to provide assistance to the nation's poorest communities to help them compete on a more level playing field. Meanwhile the Inflation Reduction Act (which is mostly about climate change) will provide $3 billion to help marginalized communities fight climate harms and another $3 billion to help them monitor industrial pollution, such as the kind reported in Lavigne's Facebook post.[35]

A few years ago, the U.S. Department of Health and Human Services released an amazing online mapping tool to help emergency response planners and public health officials identify communities that will most likely need support before, during, and after a hazardous event. The tool, called the "Social Vulnerability Index," uses census data to rank each census tract along fifteen social factors that researchers have associated with poor disaster resilience.[36] Type in a zip code and you'll find layers of candy-colored maps telling you about local income levels, racial and ethnic compositions, educational achievement, and even rates of car ownership. State and local governments are encouraged to use these maps when doing things like ordering emergency supplies or designing evacuation systems.

FEMA is building a similar database to help it identify places for special investment, though its considerations leave out the factors of race and ethnicity. At this point, a list of potential priority areas includes places like Central Appalachia, the Mississippi Delta region, South Texas, and the western coast and interior of Alaska. From a climate-and-caste perspective, it's worth noting that, with the exception of Central Appalachia, all eight entries on FEMA's list describe regions that are either disproportionately Black, Latino, or Indigenous.[37] When, in part II, we consider some specialized cases of climate resilience, from coastal restoration in the Mississippi Delta to displacement and migration in coastal Alaska, I want to keep in mind the lessons I learned in Hazira and St. James about the role of inequality in amplifying risk and the importance of inspiring leaders in protecting the people and places we love. The other thing I won't forget—which means you can't either—is the necessity of government structures that are robust, responsive, and just.

In researching this chapter, I spoke with many grassroots climate activists across the United States and in India. Again and again, one point came up, which people insisted I share: Being resilient is about more than clawing your way out of a disastrous situation without outside help. An activist friend of mine made the point well in a conversation about some folks in the bayou who had lost everything in Hurricane Ida but could not secure the FEMA trailers they were entitled to because of bureaucratic snafus. After days of angry phone calls and online wrangling, they finally scored an unassigned shipment of government tents to huddle in as they began the work of rebuilding their homes. Some in the media celebrated their pluck and down-home resilience. My friend thought otherwise. "Give me a break," she said, "People angling for tents when their homes blow away, and they call that 'resilience'? That's not resilience. That's government failure."

Put another way, climate resilience is not about battered victims being left to build back on their own. Nor is it about our leaders being munificent and kind. It is about the government being fair and making good on its part of the social contract. After two hundred years, the English writer and feminist activist Mary Wollstonecraft still has it right: "It is justice, not charity, that is wanting in the world!"[38]

chapter 5
Believing Is Seeing

One spring day, two brothers from an Iñupiat village in Alaska set out on their annual hunt for North Pacific walrus on the still-frozen plains of the Bering Sea. Dried meats and oils cured from those flippered marine mammals sustained the community year-round. The brothers were seasoned hunters, having learned the arts of navigation, tracking, and shooting from their father and uncles. Years of riding dogsleds on blank slates of ice had instilled in them an uncanny sense of direction. The brothers could find their way in almost any condition, whether smothered in snow, blinded by glare, or domed by monotonous blue skies. The contours of that world had reached further into their minds on every hunt, like needles of ice stretching over a puddle.

But this spring, the hunt would have to be different. That was because, on account of an unprecedented warming spell, the familiar plains of frozen sea had already melted. To find surviving ice floes and the congregations of walrus they attract would require traveling north several more days, a quest that would put the brothers dangerously far from home and in unfamiliar territory.

I'm hearing this story at a conference on climate displacement, migration, and relocation, cosponsored by the National Oceanic and Atmospheric Administration (NOAA) and held at the University of Hawai'i in Honolulu. A village member has flown here to testify about the environmental changes his people are seeing on the Arctic coast and how they are struggling to cope. Past louvered windows, the coconut palms are waving. But I'm focused on the ice.

The brothers drive their sleds many miles into foreign terrain, the speaker explains. They finally discover a pod of walruses that have hauled themselves on to the ice, and they are able to kill two of them. They tie the load onto their sleds and ride back toward the flat horizon from which they had come. But soon they find themselves lost. The dogs are exhausted, and the snow is kicking up. They decide to abandon the heavy carcasses and make a last effort to find their way home.

Hours later a miracle of sorts occurs: a pair of game wardens who had been following their trail intercept the brothers. Earlier, the officials had come upon their walrus carcasses, the abandonment of which is a federal crime, and set off in pursuit of the owners. (The law is meant to target ivory poachers, who saw off the tusks and leave the rest behind.) Now the wardens charge the brothers with violating federal law. The brothers, they are *elated*, the speaker says, drawing laughter from the crowd. Their lives are *saved*! Everyone returns to the village, and the legal issues are sorted out.

The speaker's account, vivid and emotionally charged, launched the audience into productive discussions about understanding and preparing for the profound climatic upheavals that are already threatening Alaska, including ice melt, coastal flooding, and the loss of native species. We agreed that most Americans hardly think about these things, and few seem to really care.

The next day, I spoke with a conference participant who said he had spoken to his dad on the phone before breakfast and had told him about the Iñupiat brothers. His father lived in Montana, my colleague told me, and for the most part was totally indifferent to climate change. He voted Republican and distrusted the long arm of government. But this time, my colleague said, his dad was absolutely riveted by the brothers' story. The father asked questions about the sleds, the means of navigation, and the worried families back home. How, the father wanted to know, would the villagers manage if the warming continues? Were they thinking of moving, changing their diet, toughing it out? Now, it wasn't like this Rocky Mountain Republican was going to join Greenpeace or buy a Prius or anything. But for half an hour over coffee, he had been *engaged*.

"Why do you think that was?" I asked.

"It was a hunting story," he said, "My dad loves to hunt. It was something he could relate to."

Despite the scientific consensus, too many Americans are skeptical about the climate crisis or doubt its urgency. Even among those who are convinced of these things, there is a hesitancy to act. People need to get engaged. The Yale Program on Climate Change Communication specializes in understanding what people in the United States know about climate change and how they weigh the risks. In its most recent national survey taken in 2021, 58 percent of the population was "alarmed" or "concerned" about climate change. The "cautious" and the "disengaged" make up 22 percent of the population, while the "doubtful" and the "dismissive" make up only 19 percent.[1]

The survey finds that Latinos and African Americans are more likely to be "alarmed" or "concerned" about climate change than whites. They are also more willing to engage in climate activism or to take climate issues into account at the polls.[2] This may be because Latinos and African Americans, as we've seen, tend to be more vulnerable to climate impacts and extreme weather.

While 58 percent for "alarmed" or "concerned" is an improvement over recent years, it's worth noting that in 1989 a Gallup poll showed that 63 percent of respondents worried "a great deal" or "a fair amount" about climate change.[3] And since then, the scientific consensus on climate change has risen substantially.

About the "alarmed" and "concerned": we know that while they need less convincing about the threat of climate change, they need guidance. Of just the "alarmed" (33 percent of all respondents), only a third are engaged in any kind of activism at all. A quarter of that group say that they rarely, if ever, discuss climate change with their friends and family.[4] The "concerned" are even less involved. The reason, I suspect, is that even those who care the most aren't sure exactly what they or their political leaders should be doing; and talking about it risks either alienating their loved ones or just bumming people out. As Kendra Chritz said when discussing her students in chapter 2, "It just seems overwhelming. To them it feels like there's nothing we can do."

About the 22 percent who are "cautious" and the "disengaged": we know that while they don't do much now, they are at least open to persuasion. They are bells that have yet to be rung. The doubters and dismissers are more hardcore, and we shouldn't worry about them. That leaves 89 percent who deserve our attention.

Decades ago, many environmentalists, including Al Gore, cautioned against discussing climate resilience because they thought it would distract the public from the work of reducing greenhouse gases. As I explained in chapter 1, this was wrong on the substance, because the essential work of reducing greenhouse gases still does little for the people suffering now. But I think they were also wrong on the communications side. In fact, they had it exactly backward. The focus on reducing greenhouse gases turned out to be vague and abstract. It was a boring, faraway topic. And because global emissions reduction requires broad international cooperation and complex regulatory structures, the issues were easy for opponents to mock and politicize.

Resilience is different. There the impacts of concern and the policies addressing them tend to be more local, immediate, and concrete. By appealing to people's near-term interest in safety and security, resilience work also helps clear a path for more abstract discussions on containing carbon emissions.

The problem is that most scientists and government leaders put too much faith in what communications experts call the "empty bucket theory." The idea is that if you shovel enough good information into someone's head, they will eventually believe what you are saying. It's related to the "expert panel theory," which says that if you get enough dentists, or physicists, or epidemiologists to agree on something, all the non-experts will nod yes and take their word for it.

You know how that goes.

What you might not know is *why* those tactics fail when politically charged topics come up. You might also be surprised to learn that our situational resistance to information and expertise is not *necessarily* a bad thing, inasmuch as it reflects important aspects of human reasoning that have helped our species adapt and thrive. We just can't let this inclination get the better of us. The solution is to *engage* people like my colleague's dad who likes to hunt. That's done by offering propositions that speak to their values and are at least somewhat consistent with their loyalties. As it happens, I think climate resilience is a perfect vehicle for this.

So, let's take the lid off our big bucket of a brain and see how it works. When it comes to complicated topics like atmospheric physics or ocean biology, all of us, let's face it, have deep knowledge holes. But that doesn't mean our buckets are empty. They are instead swimming with schools of emotion, values, and expansive group loyalties. This is good. Without these foundations for cooperative living, our species never would have arisen from six million years of flood and drought. What that means, though, is that when we evaluate a new, potentially threatening situation, our conclusions are driven not just by facts and experience, but by all those emotions, values, and loyalties—what we might collectively call a "worldview." Psychologists call this phenomenon "motivated reasoning," implying that the rational part of our thought machine is being programmed by (how to put it politely?) the *extrarational* part.

For the most part, this works fine. Many risk-based decisions we make in life—whether to carry an umbrella or when to change the tires on your car—either don't have much to do with core beliefs or are amenable to them. But every once in a while, a new situation arises that challenges some vital part of a widely shared world view. That's when things get challenging.

Dan Kahan, a professor of law and of psychology at Yale, specializes in how people evaluate public risks such as crime, disease, pollution, and the use of new technologies. He leads a multidisciplinary team of researchers with expertise in psychology, law, anthropology, communications, and political science that has conducted a range of experiments and surveys to understand how the human brain decides what to worry about.

In one experiment, Kahan and his team wanted to know how learning more facts about nanotechnology—the manipulation of materials at the molecular level to make consumer products—would affect their measure of the public risk involved. The answer was not much, at least for those who already had strong views about nuclear power or genetically modified foods. In such cases, people just imported their views on those subjects to nanotech.[5]

Another experiment, a favorite of mine, asked a group of subjects to interpret a table of numbers purporting to show the effectiveness of a new skin cream.[6] The subjects were supposed to study the numbers and say whether, according to the table, the incidence of skin rash had increased or decreased. A second group was given a similar table, but this time the researchers did something sneaky. They provided the exact same numbers but changed the labeling. In the new version the table purported to show the effectiveness of a law restricting handguns. The subjects were then asked to say whether, according to the table, the incidents of crime had increased or decreased. Researchers discovered that people whose position on gun control conflicted with the gun-control data presented to them had more difficulty reading that chart. Indeed, the higher a participant's math skills, the *more* likely it was that their political views would interfere with their ability to decipher the table on gun control. In contrast, ideological views appeared to have no effect on subjects' ability to evaluate skin cream.

When *Mother Jones* magazine reported on the study, its headline screamed, "Politics Wrecks Your Ability to Do Math."[7]

Kahan and his team then set out to tackle climate change. They interviewed a representative sample of 1,540 Americans about the level of risk associated with climate change.[8] Beforehand, subjects were given tests to measure their levels of scientific knowledge and numeracy. Subjects were also tested to determine their place on two "value scales" that social psychologists associate with risk-based decision-making: a Hierarchy-Egalitarianism scale and an Individualism-Communitarianism scale. The team found little connection between the respondents' scientific literacy/numeracy scores and their opinions on climate change.

But researchers found that concern about climate change correlated very well with the respondents' personal values. "Hierarchical Individualists," they found, rated climate change risks "much lower" than did "Egalitarian Individualists." This was so even after the researchers controlled for differences in scientific literacy and numeracy. Finally, researchers found that the area of widest disagreement on climate occurred not among subjects with the lowest scientific literacy/numeracy scores but among those with the highest.

Most people, of course, do not evaluate societal risk on their own, with or without confusing charts. Instead, they rely on people they trust to give them advice. On topics that are politically charged, research shows that in addition to objective expertise, people care a lot about cultural affinity. Thus, when test subjects were introduced to information about vaccinating schoolgirls against a sexually transmitted virus, they were much more likely to believe evidence concerning the drug's safety and its effects on sexual behavior when the physical appearance of the person delivering the message matched their cultural outlook.[9] Hierarchs, according to the study, like gray-haired men in suits. Egalitarians dig beards and denim. (Yes, in this survey all the "experts" were men.)

Thus, when we are unsure about the risk of an activity that pushes our buttons—restricting handguns, spewing carbon, getting a vaccination—we are more likely to trust people who share our worldview. In turn, those advisers, be they scientists, politicians, or the neighbors next door, are subject to the same emotional dynamics that we are. At a very basic level, human reasoning is rooted in cultural perception. Believing is seeing.

I've gotten pretty far into this chapter without mentioning the coronavirus by name, but I know you are already thinking about it. In the United States, Covid-19 whipped up the mother of all mind games. Nearly every aspect of the response became politicized. In early 2020, when the pandemic hit the United States, Americans were already dividing along party lines over the risk they said it posed to the country. After vaccines became available, Democrats embraced the jab while a big block of Republicans resisted. (As I write, the Kaiser Family Foundation reports that only 59 percent of Republicans are vaccinated, compared with 91 percent of Democrats.) Sadly, after vaccines became available, Covid deaths among Republicans appear to have surpassed those among Democrats.[10]

Whether we are talking coronavirus, crime, or climate, social media makes this effect much worse. We live in an era rife with misinformation and outright disinformation. While plenty of bad actors are pushing this trend along, it is phenomena related to motivated reasoning that make people prone to sharing and believing garbage analysis in the first place. It's a scary time.

But let's not wallow in our funk. While the dangers are real, these processes of human reasoning are as old as the species. Our sense of right and wrong has always been a complex brew of utility and passion. The distinguished biologist E. O. Wilson was one of the first to popularize the theory that human morality, in fact, evolved from emotions. Where did thinkers like Confucius, the Buddha, and Kant get their ideas? "Ethical philosophers," writes Wilson, "intuit the deontological canons of morality by consulting the emotive centers of their own hypothalamic-limbic system,"[11] Chalk one up for the lizard brain, er, the "hypothalamic-limbic system." The social psychologist, Jonathan Haidt, refers to this phenomenon as the "emotional dog and its rational tail."[12] It's how we evolved, which, by the way, is the result of millions of years of climate adaptation.

At this point you might be agreeing with Kurt Vonnegut in chapter 2. Maybe our thought machines really are "too big and untruthful to be practical." Yet the wisest among us have often viewed emotional reasoning as a feature, not a bug. Aristotle famously rejected Plato's crowning of "reason as monarch," arguing instead that reason should draw from virtues that are expressed in *emotions* like courage and fidelity.[13] David Hume, avatar of the Scottish Enlightenment, argued that "reason is, and ought only to be, the slave of the passions."[14]

Toward the end of the twentieth century, U.S. Supreme Court Justice William Brennan urged judges to be up front about their emotional motivations and argued for what he called "vital rationality" in bestowing justice.[15] Similarly, legal scholar Martha Minow and the philosopher Elizabeth Spelman have enjoined jurists to reveal and embrace the sentimental aspects of their thinking. "Passion" they write, "is no longer an attribute of 'women' or 'minorities' (or 'minority women') but a feature of all humans, including those who judge."[16]

To see why this might be a good thing, consider some developments we saw in chapters 3 and 4. Jean-Jacques Rousseau's insistence that government bear some responsibility for the consequences of natural disaster was born of indignation, as was his insight that justice be judged from the bottom up. The activism of Sharon Lavigne, Anup Karanth, and Mehul Patel sprang from the font of love, loyalty, and sometimes outrage. To be clear, I don't mean that these sentiments reinforce their commitment *after* they have reasoned through an injustice. I mean that, like Rousseau, their passions are kindled *first*. The intellectual inquiry comes later. As important as that journey is, it remains almost always under the sway of "truth-telling" emotional forces. In my involvement with environmental and social advocacy, I have seen this process again and again, and I've come to believe we would not want it any other way.

What does this mean for the climate resilience movement? It means understanding and *accepting* that advocacy is a multiarmed creature. In her recent book *Saving Us*, climate scientist Katharine Hayhoe asks us to acknowledge that "bombarding people with more data, facts, and science isn't the key to convincing others of why climate change matters and how important and urgent it is that we fix it." Instead, "study after study has shown that sharing our personal and lived experiences is far more compelling." We must, she says, "connect *who we are* to *why we care*."[17] Hayhoe, a professor at Texas Tech University who has served as a lead author on three National Climate Assessments, is also an evangelical Christian and a climate ambassador for the World Evangelical Alliance. Many evangelical Christians are known to be skeptical of climate change. Some of that skepticism might be related to factual understanding. "But by a mile," she says, "the biggest barriers are emotional and ideological."[18] So when Hayhoe speaks to conservative audiences who are skeptical about climate science, she thinks first about what she might have in common with them.

The key, she explained to a reporter at the *Washington Post*, was "showing how climate change connects to what they already care about. What they love. What matters to them."[19] In *Saving Us*, she gives this example: "Climate change disproportionately affects the poor, the hungry, the sick, the very ones the Bible instructs us to care for and love. If you belong to any major world religion—or even if you don't—this probably speaks to you."[20] Or maybe your priority is national security. Then you should know that the Department of Defense believes that climate impacts pose a growing threat to our military "capabilities, missions, and equipment, as well as those of U.S. allies."[21]

Sometimes values are braided into the solutions that accompany the identification of a problem. If you're a fan of international cooperation and group problem solving, a global effort to cut carbon pollution might seem like an existing challenge. If you distrust such tactics, your mind could be looking for ways to ignore the problem. On the other hand, some research suggests that if you are an independent-minded technologist, the possibility of expanding nuclear power or reengineering the climate might arouse your interest in the problem of global warming.[22] Finally, communications experts stress the importance of having messengers who are already known and trusted by their community. If, sporting a denim shirt and trimmed beard, the astrophysicist and self-described agnostic Neil deGrasse Tyson wandered into an evangelical Bible study group to discuss climate action, one imagines he might have a tougher time working the crowd than Dr. Hayhoe.

All this advice sounds sensible. Still, even relatively sophisticated climate advocates have trouble engaging people who say they care or who are at least persuadable on the issue. Why?

It's because the climate crisis is still mostly seen as a fuzzy abstraction. Psychology teaches that some of our strongest emotional responses are tied to tangible experience. As Hayhoe herself points out, "unlike air pollution, climate change is caused by invisible heat-trapping gases that we can't see, feel or smell."[23] Remember, too, that, because of the "thermal inertia" described in chapter 1, the people now being asked to reduce heat-trapping gases probably won't live long enough to see much of an atmospheric improvement. No wonder half of all Americans feel "hopeless" about the climate crisis.[24] The problem with being bummed out is that if you don't see a way forward, you bury your head and *check out*. Then we're cooked. How can we oppose this trend?

One way is to focus on climate-related solutions that not only tap into a wider range of social values but that are also more local and concrete. Climate resilience shows us the way. Resilience work includes a wide range of activities with the potential to appeal to many ideological tastes. Communitarians will like resilience because it involves the group in building a more robust society. Libertarians will like it because it protects their life and property. And no matter where you start, incremental progress becomes visible almost immediately. In this way, preparing for the concrete impacts of climate disruption becomes the perfect gateway for harder and more abstract discussions about greening the grid and cutting carbon emissions. Reaching *beyond* greenhouse gases thus becomes a strategy for *revisiting* it later—this time in a more successful way.

I learned this firsthand years ago, in 2010, when I was serving on President Barack Obama's Climate Change Adaptation Task Force. I led a pilot project within the U.S. Environmental Protection Agency to understand how federal research on climate impacts might be used to help flood recovery efforts in the state of Iowa.[25] In the previous two decades, that state had experienced *three* catastrophic floods, more than anyone alive had ever seen. The repeated damage had led to a federal and state effort to rebuild Iowa's devastated communities and to reduce disaster-based hazards in the region. It seemed likely that climate change was contributing to increased risks in the region. So the EPA, which had already been working with state officials and the Federal Emergency Management Administration (FEMA) to provide planning assistance, opened discussions on the second front of climate resilience. The program intended to bring together state officials, regional scientists, and community leaders for a series of discussions about the risk of regional climate impacts, such as increased flooding, heat waves, and extreme weather. EPA hoped the meetings would encourage members to identify adaptation goals and locate legal and policy tools for implementing them.

After months of work, project participants concluded that climate change projections and up-to-date hydrological data needed to be mainstreamed into Iowa's planning processes. They located many existing structures that could accommodate this. At the local level, they suggested that climate information could be integrated into comprehensive land-use plans, zoning codes and ordinances. At the state level, the group imagined folding climate resilience into the state's Smart Planning Act, which requires communities and state agencies to consider "Smart Planning Principles" when planning for future land use and development. Because planning requires community outreach and public participation, the move would also open the climate conversation not only to hundreds of local planners, but to a large swath of state residents. The project team also highlighted the potential of integrating climate scenarios into FEMA's Hazard Mitigation Assistance program, which sets standards for thousands of hazard mitigation plans throughout the country. The team's conclusion that local government and land-use decisions lie at the heart of climate change adaptation finds support in much of the planning literature, as well as in research in other fields.

The Iowa pilot project also emphasized the need for relationships among actors across many levels of government and sectors. Federal assistance through EPA and FEMA, for instance, relied heavily on the Iowa's Rebuild Iowa Office, which provided trustworthy links to scores of participating communities. Equally important was a roster of trusted organizations and individuals who had helped establish many cross-community and crosscutting partnerships. (Iowa has ninety-nine counties, many of them full of very small towns and rural communities.)

While Iowa's adaptation work is an ongoing process, the strategy of appealing to shared local values and trusted colleagues appears to be working. It was also important to keep the scale partly local and make sure the benefits, like more landscaped parks (which soak up stormwater), were tangible and near-term.

We were careful to present climate change concerns within the context of laws and programs that already existed and already reflected the public's shared values. During the pilot project meetings, participants noted that while politicians and residents were reluctant to talk about climate change in the context of pollution control, sometimes questioning the underlying science, these same people were often willing to learn about climate change and future scenarios within the context of hazard mitigation and disaster planning.

There are examples like this in many parts of the United States. In Arizona, water districts in conservative counties are using climate science to predict future groundwater supplies. In Florida, Ron DeSantis, the state's conservative governor, is committing hundreds of millions of dollars to build more stormwater pumps, move power lines underground, and otherwise prepare for the onslaught of the climate crisis. The Environmental Resilience Institute at the University of Indiana is deploying climate projections to help farmers adapt to what its website calls "environmental change." Call it what you want, for now. We can have more difficult conversations after we know each other better. Building resilience, even with baby steps, shows you care. Even if you're scared, the work makes you feel good, or at least honest and responsible, which Aristotle would say is nearly the same thing.

In their book *How to Have Impossible Conversations*, social critics Peter Boghossian and James Lindsay advise that in talking about polarizing issues, it helps to "think of every conversation as being about three conversations at once: about facts, feelings, and identity."[26] In part II of this book, I'll be surveying the landscape of climate resilience efforts in search of instructive stories and rays of hope. There will be a ton of facts—we need those if we are going to wrestle honestly with the issues—but also stories and meditations that I hope infuse the topics with heart and meaning. Facts, feelings, and identity: they are the meat that will sustain us on the journey, from brackish bayous on the Gulf of Mexico to the still-frozen plains astride the Bering Sea.

PART II
Doing Resilience

chapter 6

Moonshot on the Bayou

Psst! Listen up, I'm going to tell you a secret: What is probably the largest, most sophisticated, ecosystem-based climate-change resilience project in the world has been going on for years and most people outside the region know almost nothing about. No, it's not in Singapore, or the Netherlands, or the futuristic city of Dubai. It's in the towns and swamps of southern Louisiana—the home of Chitimacha settlers, Cajun shrimpers, and Tiana, Disney's "Frog Princess."

The best way to understand the project is to jump in a boat and see certain parts up close. That's why, every spring, as part of my class in natural resources law, I haul my students out to the Maurepas Swamp Wildlife Management Area. A forty-odd-minute drive from New Orleans, Maurepas Swamp consists of about ninety-six square miles of flooded forest consisting mainly of water tupelo and bald cypress trees, the rootlike "knees" of the latter poking out of the water like dragon's teeth. Dripping with Spanish moss, their branches shade an understory of wax myrtle, pumpkin ash, and an abundance of things that slide and crawl. Located within the "Mississippi flyway," the swamp provides rest stops for hundreds of species of transient songbirds, including warblers, cuckoos, vireos, and woodpeckers.

This morning, I have fifteen students with me, each paddling a kayak on the swamp's shimmering green water (figure 6.1). Small bits of vegetation, from duckweed to Salvinia, float along on the surface at an almost imperceptible rate. Bob Marshall, a native of Louisiana and Pulitzer Prize–winning environmental journalist, has agreed to be our guide. Marshall is a gifted storyteller, with a fine head of hair, a wide jaw, and twinkling eyes. Now semiretired, he has spent a

6.1 Loyola Law students paddling through Maurepas Swamp.

(Photo by Rob Verchick)

lifetime hunting, fishing, camping, and boating in Louisiana's coastal zone. He makes a gutsy duck gumbo, too.

For the last twenty years, his work has been dominated by the state's loss of coastland and the climate crisis. According to scientists at the National Oceanic and Atmospheric Administration (NOAA), southeastern Louisiana is sinking at one of the fastest rates of any land mass in the world. In fact, since 1932 the state is believed to have lost two thousand square miles of land.[1] The reasons are multiple and all anchored in human activity. To a visitor, Maurepas Swamp might look healthy and lush. Its life-support systems, however, are under serious stress. Louisiana's coastal resilience project, which involves using innovative technology to protect and restore vast swaths of the southern coast, could revitalize Maurepas Swamp as well as many other treasured places. But success is not guaranteed.

Back in 2014, Marshall sized up the challenge this way: "Southeastern Louisiana might best be described as a layer cake made of Jell-O, floating in a swirling Jacuzzi of steadily warming, rising water. Scientists and engineers must prevent the Jell-O from melting—while having no access to the Jacuzzi controls."[2] The state's plan to restore the coast is audacious, what Marshall has called a once-in-a-lifetime moonshot.[3]

Marshall dips his blade into the water, and a ribbon of kayaks—in bright shades of blue and red and yellow—follows behind. "Swamps provide some of the best storm protection around," he says. Because of their ability to dampen storm surge, wetlands are sometimes called "horizontal levees."

But there are many ways to use a swamp. "Cypress trees are fabulous wood for houses and boats because the grain is so tight," he says. "So, when Europeans got here, they cut them down and built houses and boats. And they kept cutting them down." My house, which drowned in Katrina and survived the tell the tale, was built of old-growth cypress. (More about Katrina and that house in the next chapter.)

A largemouth bass breaks the surface with a splash. There's a lot of life in this forest, from bass to otters to ribbon snakes. And, of course, alligators. Which is why I insist: no marshmallows . . . and no dogs.

"Swamps like these are all dying, but they're still very beautiful." Marshall says. "Part of the plan under construction is for a big pipeline to bring five thousand cubic feet of river water a second into this area. And when the water pours into here, everything will change." That pipeline of water—and the sediment—it carries with it, is the key to invigorating the state's collapsing swamps and marshes. Combined with other engineered conduits, it will form the booster rocket for Louisiana's moonshot toward resilience.

—◆—

Of all the threats associated with the climate crisis, sea level rise is the one most people think of first. Climate breakdown has accelerated global sea level rise to the fastest rate in more than three thousand years. In the last hundred years, tides around the world rose by ten to twelve inches. NOAA scientists predict that ocean levels will rise at least that much in the next thirty years. Even if the world takes swift action to cut carbon emissions, that trajectory is locked in. Looking ahead to the end of the century, limiting the amount of heat-trapping gas could mean the difference between sea levels stabilizing at about two feet above the historical average or surging by almost eight feet. Rising seas, along with stronger and more frequent storms, endanger all of the nation's shorelines, which touch three oceans, the Gulf of Mexico, the Caribbean Sea, and the Great Lakes. More than 130 million Americans, or about 42 percent of the population, live near coastal shorelines.[4]

Sea level rise, on its own, threatens roughly $1 trillion in national wealth held in coastal real estate. This will require many owners to harden perimeters, elevate structures, or simply abandon their land. In the "high-emission" scenario, federal experts see a future where up to half a trillion dollars' worth of real estate is permanently underwater by 2100. There is also a phenomenal amount of built infrastructure on the coast: roads, bridges, tunnels, pipelines, power stations, refineries, and more. The entire country depends on coastal seaports for access to goods and services. Together, these ports handle 99 percent of all overseas trade.[5]

National security is also at stake. In the last chapter, we saw that dozens of military installations in southern Virginia, including Naval Station Norfolk, were being seriously threatened by rising seas and stronger storms and that the Department of Defense is moving quickly to armor shoreline, elevate buildings, and harden infrastructure. Vulnerabilities like these exist wherever land and tide meet. Consider also how much the military's success depends on improvements made in adjoining communities where so many sailors, soldiers, and other employees live. For instance, in Norfolk, increased flooding threatens access routes, historical neighborhoods, and personal and commercial real estate. The city is investing in hundreds of millions of dollars to improve storm water pipes, flood walls, tide gates, and pumping stations. It is also stabilizing the coastal edge with natural materials like rock, sand, and shells, creating a barrier referred to as "living shoreline" to reduce erosion and improve water quality. While some neighborhoods have been identified for priority protection, others, where containment is more difficult, may eventually be abandoned.[6]

Protecting coastal communities will be very expensive. Government experts have no reliable estimate of how much money that will require. But they know the cost of doing nothing is much higher. According to the National Climate Assessment, damage to coastal land could total $3.6 trillion through 2100, compared to $820 billion where cost-effective adaptation measures are implemented.[7]

As a law professor, I'm particularly drawn to the blizzard of lawsuits that are going to follow from conflicts over who is responsible for protecting what. My students and I unpack such puzzles regularly. May an insurance company recover against a city on the grounds that the city's failure to protect against climate-related flooding resulted in more insurance claims? (I'm skeptical of the theory, but in 2014, Farmers Insurance brought such a claim against the City of Chicago, only to withdraw it weeks later.) May residents of low-lying homes require the government to elevate the roads so that they can continue to access their property by car? (Yes! So suggested a court in Florida.) May owners of a beachfront home require the state to compensate them for building a protective sand dune on their property, thus ruining their view but protecting the home (and neighboring homes) from being pulverized by violent storms? (Yes, again, although the value of the protection must be deducted from the value of the lost view. In this way, homeowners in Harvey Cedars, New Jersey, were able to recover one dollar in compensation.)[8]

One of the most promising ways of protecting the coast involves nature itself. The coasts are also home to important natural systems, such as beaches, intertidal zones, oyster and coral reefs, salt marshes, estuaries, and deltas. These systems are essential to coastal erosion because they help to dampen waves and reduce erosion. There has also been interest in hybrid or nature-based features such as living shorelines to protect coastal communities.[9]

Indeed, swamps and marsh provide one of the great weapons against global warming. They reduce greenhouse gas emissions by absorbing carbon dioxide in the same way that terrestrial forests do. And they provide the first line of defense against erosion, waves, flooding, and storm surge. Sadly, these aqueous ramparts are losing their mojo. Scientists estimate that between 2004 and 2009, coastal wetlands in the United States have eroded or been purposely destroyed at an average rate of 80,160 acres per year, with 71 percent of that damage occurring in the Gulf of Mexico.[10]

Driving on the Earhart Expressway on the way from New Orleans to Lake Maurepas, you might notice a standard road marker with the highway number (3169) floating over the state's chunky bootlike silhouette. Such an expansive, lush area for two and a half million people, you might think. A "Sportsman's Paradise," as the license plates say. But looks are deceiving. Maps maintained by the U.S. Geographical Survey tell a more complicated story, especially the maps that exclude all land and vegetation that is inundated by water.

Bob Marshall had made this point to my students in a classroom lecture a few days before our paddling trip. "You might think you live in a big area," he said, referring to the state's coastal region, "but it's mostly wetlands and we're still losing them at a dramatic rate. And when they're all gone, we'll just be two little fingers of land sticking out into the Gulf of Mexico, barely speed bumps for hurricane storm surge."

He told the class that it might be possible to slow this destruction, but that first you had to know how this miraculous habitat had developed over many thousands of years, how it had become the largest expanse of coastal wetlands in the contiguous United States.

"As I explain," he said "ask yourself these questions: What would happen to the wallets of most people if the nation lost 50 percent of its refining capacity, 20 percent of its natural gas supply, 90 percent of its offshore energy supply? If farmers all up and down the Mississippi Valley couldn't ship their grain? If 31 percent of states lost their access to foreign ports? That's what's at risk here."

He recites the history of the coast as if it were the genealogy of a beloved family elder, but who took many millennia and an unimaginable amount of mud to make her debut. Six thousand years ago, you see, southeastern Louisiana didn't exist. The land that would become New Orleans was a seabed, the playground of snails, squids, and bivalves. Lake Pontchartrain, the enormous estuary that today forms the city's northern border, was just a stretch of dry land perforated by a network of rivers and streams. These waters were fed by what we now call the Mississippi River, a channel of water that was already tens of millions of years old.[11] The northern glaciers, left over from the last ice age, were still melting, but more slowly than before. The world's oceans, which for thousands of years had been rising by three to seven feet per century, were beginning to settle at the level we know today. Fresh water tumbling from the Mississippi onto the southern plain began to form a small bay, the beginnings of Lake Pontchartrain.

About four thousand years ago, the Mississippi River shifted toward what is now New Orleans, spilling mountains of sediment and building dry land. A few hundred years after that, the river shifted again, shoring up the St. Bernard delta and nearly enclosing Pontchartrain from the Gulf of Mexico (hence, the misnomer, "Lake"). The Mississippi and its distributaries dumped sediment into the shallow gulf waters, building more and more land—broad ridges, peninsulas, barrier islands—wherever the spigots were pointed. About two thousand years ago, the river shifted west, robbing the eastern deltas of mud. Erosion and land loss ensued, whittling out, over the next thousand years, the Chandeleur Islands and other barrier chains. Eventually, the spigots swung east again and slowly coughed up the Plaquemines Delta, that skinny appendage stretching southeast that looks like a bird's foot.[12]

As these events unfolded, human settlements were expanding in the region. Between 1730 and 1350 BCE, an Indigenous population known as the Poverty Point society occupied the Lower Mississippi Valley and southeastern Gulf coast. Recognized by archeologists as one of the continent's oldest complex cultures, its members were expert mound builders, constructing hundreds of earthen monuments across the Southeast. The largest, a seventy-two-foot-tall mound of enormous concentric half-circles, is recognized as a UNESCO World Heritage Site, alongside Egypt's Great Pyramids and the Taj Mahal.[13]

Now think about the river that carved this valley, the ancient Mississippi, from which towering sauropods, crocodilians, and mastodons have supped. It drains everything between the Rocky and the Appalachian Mountains. That's about a third of the entire continent. "Every drop of rain that falls and every flake of snow that melts," Marshall said, his eyes catching a spark, "runs downhill through this valley picking up dirt and carrying it into creeks to streams and to rivers." That amounts to millions of cubic feet of sediment. In the spring, before modern levees and dams, there was so much rain and snowmelt that all that water and dirt could not be contained. When the river overflowed, it swept over the deltas leaving behind smears of sand and clay, which gradually rose higher than the floodwaters could climb. The water would then have to go somewhere else, scouring another channel or jumping another bank. That is why, over the span of thousands of years, ancient rivers like the Mississippi swing back and forth, like the slow-motion whips of an unattended garden hose.

By the eighteenth century—to skip briskly ahead—the distributary channel of the river stretched from the present-day state of Texas to Mississippi. By then, the region was populated by a variety of Indigenous cultures, as well as by people from Europe, Asia, and Africa. The estuary supporting them contained more than six thousand square miles. It was, to use Marshall's phrase, "the Amazon of North America."

Today our Amazon is withering and falling in on itself. As a case in point, Marshall called our attention to Bayou Terre aux Bouefs, a well-known distributary in the southeast corner of the state that slides into Breton Sound. Its banks are surrounded almost entirely by open water. In French, the name means something like "River of the Land of Wild Cattle." Marshall scratched his head. "Now why do you think the French named this bayou for wild cattle?" "Because," he said, "they saw *buffalo* on either side. Now, buffalo don't have webbed hooves. They don't live in *water*. They're *prairie* critters." The point is that buffalo—or American bison, if you prefer—had for centuries been grazing on subtropical prairies jacked up by river sediment. But no longer.

The cause of this transformation was a combination of natural processes and human intervention. The first thing to understand, according to Marshall, is that the land underlying the Louisiana coast is more dynamic than most parts of the country. Made of saturated layers of sand and clay, it's basically a big sponge resting on a hard layer of rock. "You see, gravity presses down on the muddy sponge," he said, pushing one open hand against the other, "and squeezes the sponge against a hard layer of rock underneath." In the dry season, the sponge compresses and the water is forced out. In former times, springtime floods would replenish the sponge and cover it with a new layer of silt.

The system worked well, but the floods endangered human settlement. So, almost as soon as people began living on the delta, they began to manipulate the environment to build resilience. Indigenous peoples engineered mounded highlands and armored stream banks with mud, rocks, and shells—a forerunner of the hybrid "living shorelines" we use today. European settlers designed earthen levees and dredged flood basins. These innovations changed the natural systems somewhat, but for the most part, the coast's muddy sponge worked as it always had—squishing water out in fall and winter, then slurping it back up in the spring and summer.

Then came the Great Mississippi River Flood of 1927. The winter of 1926 had covered the Midwest with record snowfall. The spring of 1927 brought record rainfall. The Mississippi River turned violent, leaping its banks and busting through levees at will. Twenty-seven thousand square miles of land, from southern Illinois to the Mississippi Delta, were drowned. Hundreds of people died; hundreds of thousands were left homeless. "So Congress said, 'That's it,' and told the Army Corps of Engineers to put that monster in a straitjacket so it would never happen again," Marshall said. "Then they built the world's greatest river levee system on the Lower Mississippi."[14]

A breathtaking feat of science and engineering, the levee system caused two new problems. First, it halted the process of land-building by preventing silt from pouring over the banks onto the surrounding prairie. Second, the levee system deprived the spongy layers of clay and silt from the water that kept it fat:

lands that now supported highways, buildings, and railroads began sinking even faster. "If you've lived here for any length of time," explained Marshall, "you know that our streets are like carnival rides, going up and down. We have potholes the size of houses. Our houses lean one way and the other. That's because we impounded the delta and covered the surface with impervious materials that prevent our subtropical rains from recharging the sponge."

"By the way," he added, "it's not just us. Every major delta in the world is in serious trouble." He's right. Consider the Egyptian city of Alexandria, which has for millennia thrived on the Nile Delta. Surrounded on three sides by the Mediterranean Sea and backed by a lake, this fabled trading hub has recently endured terrible storms and flooding that are threatening poorer neighborhoods and drowning archaeological sites. Bangladesh, situated mostly on the Ganges Delta, is sinking nearly four centimeters (about 1.6 inches) a year. Increased cyclones and flooding have destroyed millions of livelihoods in recent years, putting the future of the nation's entire economy into doubt. In Vietnam's Mekong Delta, experts fear that up to twelve million people may be forced to retreat from rising seas in the next fifty years.[15]

Back in Louisiana, NASA is deploying high-tech airborne systems to analyze the parts of the delta that are growing and shrinking. Scientists from NASA and several universities hope to use this information to build a computer model that will help other countries identify the parts of their dwindling deltas that would respond best to intervention.[16] Before we get to Louisiana's restoration efforts, however, I need to introduce a final variable to the equation of the state's shrinking wetlands. We know the surrounding seas are now rising a few millimeters a year. We know that the channelization of the river is preventing new land from gradually being built. Still, in the span of one human life, this state has lost two thousand square miles of land. A delta cut off from river flow typically takes centuries, if not millennia to fall apart. How did Louisiana lose so much land so quickly?

Answer: the fossil fuel industry.

In the beginning of the last century, Louisiana's coastal zone was, as it is today, nearly all privately owned (various Indigenous claims having gone unacknowledged or ignored.) Then there was not much development. For the most part, life on the coast revolved around hunting, trapping, and fishing. That changed in 1901 when Jules Clement, a farmer in Evangeline Parish, near the center of the state, noticed bubbles of methane popping in his rice fields. Weeks later, he would erect the state's first oil drill. The first of many oil booms would soon follow. To date, more than 220,000 oil and gas wells have been drilled in Louisiana.

Of those, fifty thousand were placed in Louisiana's coastal zone. Most were accessed by canals, up to two hundred feet wide and fifteen feet deep. Combined with waterways cut for shipping, the state now has about fifteen thousand miles of canals. The channels allowed salt water into the freshwater estuaries, which killed vegetation and destroyed habitat. As the canal makers dredged, they dumped the sediment on either bank of the canal, making berms called "spoil levees" on either side. These structures, which can be as high as fifteen feet, block any lateral flow of water and sediment.

"So now you've got fifteen thousand miles of canals, thirty thousand miles of spoil levees, and this amazing expanse of wetlands—the Amazon of North America—falling apart like a moth-eaten sweater," says Marshall. So, there you have it: citrus orchards, cattle ranches, fields of cane and cotton, churchyards, fish camps, ball fields, and ancient middens—all boxed up in Davy Jones's Locker.

Maurepas Swamp has not escaped these challenges. These wetlands are scored with canals cut by oil companies and, before that, logging outfits. Old growth cypress is rare here, although earlier in the day we did see one ancient straggler, marked with a brass plaque tracing its history back to 1803: "Louisiana Purchase Cypress Legacy," it says.

The swamp's regenerative capacity has been hampered by the resulting salt-water intrusion. Levees along the Mississippi River have starved the swamp of spring floods for almost two hundred years. According to researchers at the U.S. Geological Survey, the Maurepas region lost a quarter of its land from 1932 to 1990. Without intervention, it is estimated that half of what remains will be gone by 2050.[17] That would be devastating to much of the wildlife here, including the anadromous Gulf sturgeon, a federally listed threatened species that is believed to spawn in this area. Or the prothonotary warbler, another declining species, whose yellow robes and smoke-blue wings I spy through a veil of Spanish moss. These birds, named after brightly dressed papal clerks, are some of the few warblers known to nest in the holes of standing dead trees.

Our convoy of kayaks slides out of the closed canopy and into a bright rectangular field of shimmering water. A red-eared turtle plops into the water as we pass. Along the left shore are a series of perpendicular channels, each straight as the shrubbery at Versailles. Their spoil banks, only a few yards in width, are high and dry, providing real estate for live oaks, which don't like getting their feet wet. These leafy corridors are picturesque, but they are misplaced—barricading the natural circulation of fresh water. "Like bones in a graveyard," Marshall says.

My students and I corral our boats into a circle. Fossil fuel production is, again, on Marshall's mind. He's going to describe how it crawled off the Louisiana shoreline and learned to swim. "We taught the world how to get oil and gas from water offshore," he says referring to the early decades of the last century when fuel production in Louisiana's marshes and bayous began to rival that of Texas's famous Spindletop oilfield. Local shrimpers and trappers soon found jobs as roughnecks and drillers, working for corporations like the California Company and Shell Oil. Later in 1947, a consortium of companies led by Kerr-McGee and Phillips Petroleum would drop the first well that could not be seen from shore, marking a new phase in global fossil fuel development.[18]

"Eventually," says Marshall, "we put 4,500 rigs off our coast, the largest concentration of offshore oil and gas structures on the planet. And how do you get all that wealth ashore? About ten thousand miles of pipeline cross our coast and to other parts of the country. Why? Because half the nation's refining capacity is here." According to research conducted by university scientists, the government, and the oil and gas industry itself, this fossil-fuel development is responsible for 36–60 percent of all coastal land lost in Louisiana since the 1930s. In some places, it's more like 90 percent.[19]

We leave the open rectangle and paddle into another canopied bayou. Along the way someone spots our first alligator of the day. This one's a juvenile, no more than a couple of feet long, lounging on a floating log. You can tell the young ones (notwithstanding their size) by the bright yellow stripes on their tails. A student to my right spots another crowd-pleaser: a six-spotted fishing spider nestled in some cypress roots. This patient creature will lay for hours in the water, waiting for a tadpole or fish of just the right size. Then in a flash, she will scamper across the water's surface, seize her prey, and immobilize it with a venomous bite. It's best not to touch them, I say.

At this point, I suggest to my students that it might be time to inject some legal knowledge into our proceedings. Gators and bugs are fine, but I have a syllabus to keep up with—as well as a dean to explain things to. We sweep our boats toward a nearby bank in the shade of some cypress and red maple. A couple of paddlers take hold of a tapered cypress knee thrusting up near the bank. The rest draw closer in parallel, each holding on to the webbing of their neighbor's deck. We form a raft of sorts. I take a swig from my poly bottle, then start explaining.

While much of the coast is privately owned, I say, the law acknowledges that these magnificent ecosystems are something we humans all rely on, as well as other members of the biotic community. Louisiana has several agencies that are supposed to protect the people's treasure from being looted. In 1972, the U.S. Congress passed a statute called the Clean Water Act, which, in Section 404, requires permits from the Army Corps of Engineers for anyone wishing to plow through coastal wetlands. Of the thousands of miles of the pipelines and canals that have ruptured the coast since that time, nearly all are controlled by such permits; they typically require spoil banks to be removed and canals to be back-filled when their use is abandoned. This rarely, if ever happens. To give one example, a recent report published by the U.S. Government Accountability Office found that oil and gas producers have been allowed to abandon 97 per-cent of offshore pipelines in the gulf without incurring any penalties.[20] No state or federal official has held these companies accountable for the requirements placed on most permits.

There are other laws too. There is the National Environmental Policy Act of 1969, or NEPA, that requires federal agencies to review the environmental impacts of federal activities. Thus, when the Department of Interior rolls out its offshore oil-and-gas leasing programs it is supposed to scrutinize potential harms related to spills, leaks, pipeline expansion, and the dredging of canals. This review now also includes assessment of an activity's greenhouse gas emis-sions and its effect on climate resilience.[21] Despite such investigation, the coast continues to be chewed up by industrial development and remains seriously damaged by the 2010 Deepwater Horizon explosion and oil spill, a monumental disaster whose impacts were grossly underestimated in the original approval processes.[22]

There is also state tort law—the flexible body of law meant to compensate individuals when someone else has wrongly caused them injury. Over the years, oil and gas companies have been sued countless times for the damage their activ-ities in the marsh have caused, from polluting property to destroying wildlife habitat to dismantling the natural storm barriers that protect the coastal par-ishes. Some tort lawsuits succeed. But generally, the big ones don't. Twice, the levee board responsible for protecting New Orleans has sued oil and gas compa-nies for destroying vast stretches of wetlands important to hurricane protection, and twice, a federal court has found that the state's tort law is not expansive enough to address such attenuated and large-scale injuries.[23]

Sometimes, I explain, gesturing toward a nearby canal, you will read headlines in newspapers and magazines that talk about "vanishing" wetlands and the state's "disappearing" coast, as if this were some gently unfolding, overproduced David Copperfield trick.[24] No: our wetlands were intentionally and violently destroyed—with explosives, seismic drills, and multi-ton amphibious excavators. To this day, that is still happening.

Understanding the connection between the fossil fuel industry and Louisiana's decimated coast is important, I say, for two reasons. First, the fossil fuel economy not only contributes to greenhouse gas emissions, but it also undermines our ability to cope with amplified storms and floods. That means that cutting carbon and building resilience both require the same thing: responsible limits on the industry's damaging activities. Second, because oil and gas development has directly contributed to and benefited from the destruction of these wetlands, the industry has a duty to help repair the damage on a scale commensurate with the harm it caused. *Responsibility* and *accountability*: without these ingredients, our swamps and marshes cannot survive.

Marshall pulls his yellow boat out into the water to face us. "Louisiana is in big trouble," he tells the group. "People think we're all a bunch of crazies. As long as we've got parades and beads and football and crawfish, we're happy. *Laissez les bons temps rouler!*" he sings—Let the good times roll! "Maybe," he shrugs. "But they should still care. Here's why . . ." He rattles off more facts on coastal benefits, this time about seafood: The gulf has the most productive estuary south of Alaska; 75 percent of finfish and shellfish depend on it; half the nation's shrimp is landed on state shores; 40 percent of the oysters and 35 percent of the blue crab. Indeed, an influential study conducted in 2010 concluded that the Mississippi River Delta provided $12 to $47 billion in benefits to people each year. If this "natural capital" were treated like an economic asset, its authors said, its annual economic benefit would total between $330 billion to $1.3 trillion.[25]

Protecting that asset is a main concern of the state government's coastal resilience strategy. That strategy is really a combination of projects meant to keep people and property safe while allowing for the services provided by coastal ecosystems to function and in some places thrive. Its strategy follows the motto, "Resist. Adjust. Retreat." With an emphasis on "Resist." The best-known feature, which we've already touched on, is the Greater New Orleans levee complex, officially now called the Hurricane Storm Damage Risk Reduction System. The Army Corps of Engineers rebuilt this interlocking colossus of gates, levees, floodwalls, and pump stations in response to the levee failures after Hurricane Katrina. Stronger and more resilient than the city's ramparts have ever been, the system, which cost $14.5 billion, is intended to protect against a storm that has a one percent chance of occurring in any given year (sometimes called a "one-hundred-year storm"). For those living inside the system's 133-mile perimeter, home values are stable and flood insurance basically remains affordable. For those living outside the boundary, as we'll see, life gets more complicated.

The new levee complex, which boasts the largest water pumping system in the world, is a technological marvel. But even more revolutionary is the state's Coastal Master Plan—Louisiana's moonshot on the bayou. First announced in 2007, Louisiana's Coastal Master Plan sets an ambitious path to respond to the loss of the state's coastal land, bursting with scores of proposed projects to build resilience. The plan requires "the best available science and engineering" to identify and sequence the top projects. It also promises to engage communities and other interested parties in the planning process. Implementation of the plan was envisioned to take place over fifty years at a cost of $50 billion but is estimated to reduce damage over that time by $150 billion.[26] Which is to say that, if these numbers are right, the plan would pay for itself three times over.

The document follows a syllabus that our octopus teacher in chapter 3 could have drafted. First, it is *flexible*. By its terms, the master plan must be completely reassessed and updated at regular intervals (in practice, every five to seven years) and then approved by the legislature. This is because scientific understanding, engineering technology, and political and economic behavior are always changing. The worst thing you can do is lock yourself into a course of change that later becomes misguided or irrelevant. The most recent version of the plan, as I write, was approved in 2017. The next version is expected to be approved in 2023. We'll look at some important changes in a moment.

Second, Louisiana's master plan is *forward-looking*. Unlike many resilience measures, this restoration plan incorporates climate change forecasts into several key variables, including sea level rise, storm activity, and river flow. This is accompanied by one of the world's most sophisticated monitoring programs to check the forecasts against real-world developments. As scientists learn more about the effect of warming on elements like fish migration, ocean currents, and glacier melt, the information can be folded into updated versions of the plan.

Finally, the plan is *fit-for-scale*. Before 2005, state efforts related to coastal protection and restoration were handled by a turbid sea of state and local agencies, each with limited budgets and almost no outside coordination. There was no CEO-style cortex, no hublike brain. In 2005, following the devastation of Hurricanes Katrina and Rita, the federal government agreed to provide recovery assistance and funding, but insisted that Louisiana place its coastal activities under the supervision of a central authority that would represent the state and be accountable for aid received. It was also important the authority to have a single, coordinated plan with clear goals and achievable objectives.

The result is the state's Coastal Protection and Restoration Authority, or CPRA. The CPRA is charged with drafting and implementing the master plan. Any major infrastructure or activity planned in the coastal zone must be reviewed and approved by this agency as being consistent with the plan's goals and objectives.[27] All of this work is supposed to include a lot of collaboration with other government entities and interested members of the public—a way of taking advantage of the state's "sprawling brains and rubber arms" we talked about in chapter 3. By law, the CPRA must consult regularly with "coastal partners," recruited from universities, state agencies, industry, and nonprofit organizations. Since its beginnings in 2007, that agency has hosted more than a hundred public meetings. Parts of the plan have also been reviewed by the federal government under NEPA, a process involving an Environmental Impact Statement and more public hearings.[28]

Of the plan's $50 billion budget, about $19 billion will go to structural defenses for communities along the Gulf Coast. The marquee item is a second levee system, now under construction, that aims to protect 150,000 residents living in the parishes of Terrebonne and LaFourche, about sixty miles southwest of New Orleans. Known as the Morganza-to-the-Gulf Hurricane Protection System, the $3 billion project would provide the same level of protection that New Orleanians enjoy. It could be completed as early as 2035.[29]

Twenty-five billion dollars is allocated to try to hold onto some of the wetlands that the state has left.[30] Engineers can't bring back all of the demolished two thousand square miles, but they can strengthen some of what's left and rebuild other parts by using a process that Marshall calls "pumping and dumping," that is, pumping dredged sediment from nearby rivers and bays and dumping it into cratered basins on the delta. The problem is that on account of continuing subsidence (remember the shriveling sponge), the pumping and dumping must be repeated every twenty years or so.

A more sustainable and exotic solution involves sediment diversions. This is a way of letting the natural forces of the river build land the old-fashioned way. The engineering is complicated, but the basic idea is to insert floodgates at key points on river levees and occasionally let the monster jump out of its box. When the river is swollen and thick with sediment, the gates are opened to allow mud to flow over the delta and rebuild land.

The master plan envisions a dozen of such diversions on both the Atchafalaya River (in south central Louisiana) and the Mississippi. One of these diversions will bring life-sustaining fresh water and sediment from the Mississippi into Maurepas Swamp, restoring roughly seventy square miles of forested wetlands. Begun in 2017, the project envisions three years of preparatory engineering, design, and land acquisition, and another four years of construction.

The two largest of these projects—known as the Mid-Barataria and Mid-Breton Diversions—are planned near the mouth of the Mississippi River between New Orleans and the terminal Bird's Foot Delta in the southwestern part of the state. Projects on this scale are not just new to Louisiana but to coastal communities worldwide. Universities and businesses are inventing technology that can be exported across the globe, potentially improving the lives of millions of people. The state's modeling approach, designed specifically to address the needs of the planning process and provide quantitative comparisons of options for action, is admired around the world. All that has been extremely good for business, with the state's high-tech water management sector projected to grow more than 23 percent in the next ten years.[31]

Still, there are disappointments. In 2012, after consulting the best science at the time, the plan's modeling team had concluded that over fifty years these restoration efforts would build more land than would be lost. Five years later, newer models gave us what Marshall calls "the big raspberry." The new models, now factored into the 2017 version of the plan, predict that while restoration efforts will build eight hundred square miles of land by 2067 and continue building more beyond that, the coast will simultaneously shed two thousand square miles of land. That's a net *loss* of twelve hundred square miles, an amount roughly the size of Delaware. What changed? The forecasts on sea level rise. Based on climbing greenhouse gas emissions, faster rates of ice melt, and resulting rising tides, the *worst-case* scenario for future sea level rise in the 2012 plan became the *best-case* scenario in the 2017 plan.[32]

From this revelation, two hard truths follow. First, Louisiana's moonshot can extend life, but not conquer mortality. This should not really surprise anyone. Understood honestly, climate resilience in any context is basically a way to buy time. Ideally, that time can be used to aggressively ratchet down greenhouse gas emissions and remove further stress from the system. Even if the master plan increased total land mass over the next century, there would inevitably be a point at which the numbers would turn negative. It just depends how far into the future you look. According to research conducted by climate scientist James Hansen, within the next *two hundred* years, sea levels could rise as much as twenty feet.[33] That would drown most coastal settlements on the planet, from Los Angeles to Copenhagen to Kolkata.

If this sounds defeatist to you, let me surprise you by saying that, instead, I hear a ring of honesty and liberation. To make a comparison: no individual person lives forever. Still, we pursue satisfaction and joy in our hour of struts and frets upon the stage. And we make deliberate and sometimes heroic efforts to capture that full hour, always scheming for a few seconds more. That's good. I want my hour to mean more than "a tale told by an idiot," as Shakespeare's Macbeth said. The same goes for the lifespan of the Mississippi River Delta and its varied peoples. Bestowing another fifty or a hundred years of capability on a people is a precious gift. It protects livelihoods, allows opportunities to plan, and, importantly, buys time for future greenhouse gas reductions so that the damaging impacts will be further off and less severe. Also, don't forget that over fifty years this plan still saves three times more money than it costs.

The second hard truth is that no matter what *Star Trek's* Captain Picard thinks, resistance is—at least at some point—futile. That doesn't mean you don't try it. But you must also be ready to adjust and retreat. This is exactly what the 2017 Master Plan now advocates. It earmarks $6 billion to help people elevate homes, waterproof equipment, or move. Most of the money is focused on residents, but some businesses are eligible too. Everything is voluntary. Further, there are special provisions to make it easier for low-income families to take advantage of the programs, though many say it could be more generous. Under the plan, more than twenty-two thousand homes would be elevated to survive flood depths of up to fourteen feet and more than two thousand homes (those subject to even higher floods) would be targeted for voluntary buyouts.[34]

At this point you may be drumming your fingers on the table, wondering about a question I have so far managed to avoid: *How is this $50-billion moonshot getting paid for?* The answer is not completely satisfying. The federal government has so far committed about $10 billion through appropriations targeting water works and disaster recovery.[35] In compensation for the Deepwater Horizon explosion and oil spill, Louisiana was awarded annual payments of about $1 billion for a period of fifteen years. During the George W. Bush administration, four Gulf states brokered a deal for an enhanced portion of federal royalties from offshore fossil-fuel production. The arrangement provides Louisiana a limited, but dedicated, stream of revenue earmarked for coastal protection and restoration. In addition, the infrastructure bill, which Congress passed in 2021, is expected to pump another $3 billion or so into the state's restoration efforts.

One obvious problem is that the financial commitments thus far fall way short of what is needed. Louisiana politicians have been angling for more federal support. But now that states like Texas, Florida, and New York are also contemplating expensive flood-protection projects, the competition is getting keener. If Louisiana can't raise the funds, it will have to scale back the projects.

A second problem, if you're thinking on a national scale, is that this mechanism is too idiosyncratic to be replicated. The largest source of funding is the legal settlement from one of the nation's most ruinous environmental disasters—not exactly a game plan for the future. That said, in the absence of federal action, several states and municipalities have sued fossil fuel companies for their major role in the climate crisis. Often using theories of state tort law, plaintiffs hope for multibillion-dollar damage awards that would then be used to adapt to future threats.[36]

I welcome these efforts, but mainly as a way of holding polluters accountable for their destructive actions. As a national mechanism for funding resilience, they're, again, too idiosyncratic, relying on unique fact patterns, the choice of judges and jurors, and access to lawyers, scientists, and money. Did I mention that, so far, no lawsuit like this has yet succeeded? A better way, I think, would be to hold fossil fuel producers accountable by taxing carbon and then using the proceeds to help communities build long-term climate resilience.

To the swamp paddler, lettuce mats are a scourge. I mentioned earlier the sprinkles of vegetation that paddlers find on the bayou, the duckweed and Salvinia fern. Navigating the bayou can be a little like paddling through lentil soup. But here we've been stopped in our tracks by a bank-to-bank installation of water lettuce. Water lettuce is a kind of floating plant about the size of an outstretched hand. It's dull green with thick hairy leaves. In large congregations, they resemble an undulous rubber carpet, which locals call a "lettuce mat." Marshall pulls a lettuce head from the water, exposing a system of long feathery roots. Water lettuce, which may or may not be native to the continent, is a nuisance, Marshall tells us. It's known for depriving the water of light and oxygen and degrading the habitat. It also clogs waterways. Although we could push through, we decide that, given the time, we will return to the launch. With levees and channels and lettuce mats, there's a lot to ponder. "In addition, we have this other problem," Marshall deadpans. "It's called 'climate change.' It's very popular among people who believe in science."

Marshall pokes fun at climate deniers, but for reasons stated in in the previous chapter, I try not to be as judgy. Indeed, in the last thirty years Louisiana's journey toward resilience has brought its leaders and citizenry a lot closer to reality. Consider that in 1993, when the state finally began to tackle this issue, human-induced climate change was an unspeakable thought. That year, the state's task force on coastal restoration, whose members were appointed by Governor Edwin Edwards (a Democrat), would call out "global sea level rise" in its inaugural report but insist—against the report's own cited evidence—that the phenomenon was a "natural" process.[37] Gradually, as the state began lobbying the federal government for serious money, the masquerade ended. By the time Governor Bobby Jindal—an erstwhile hero of the Republican movement and climate-change skeptic—would put his name on the cover of the plan's 2012 report, the document would specifically link projected increases in sea level rise to greenhouse gas emissions caused by human activity.[38] That version and the 2017 version—whose warnings about human-caused climate change are even more pointed—were both eagerly embraced by the state's conservative legislature.

Now fast forward to 2020: under the leadership of Governor John Bel Edwards, a conservative Democrat, Louisiana became the second of only two states in the South to pledge to eliminate all net greenhouse gas emissions by 2050.[39] The following year, a new state task force—whose members represented a range of community and industrial interests—*unanimously* accepted a plan that would expand renewable energy, electrify parts of the manufacturing sector, and increase energy efficiency for low-income housing. I don't know how much of this the politicians will actualize. But take a moment to marvel that such a thing happened at all. Louisiana—a red state in the Deep South, the cradle of offshore oil production, which spews more carbon per capita than nearly any other state—has pledged to eliminate all net greenhouse gas emissions in less than thirty years.

In the interest of full disclosure, I am one of the members who served on the task force that adopted that net-zero proposal. The motives of individual members were multidimensional and sometimes fluid. But there was one goal that when mentioned would always unite the group, across lines of political affiliation, expertise, geography, income, or race: *saving the coast from annihilation*. The crystallization of that goal—backed by decades of scientific inquiry specific to the state—would never have existed without the commitment of that earlier task force in 1993 to recognize the need for ecological resilience. For me, this is a classic example of "Believing is seeing," which we explored in the previous chapter. Tap into people's emotional ties and concrete interests and let science seep the through the cracks.

On the paddle back, Marshall and I have no more lectures. Our tour is quieter and more soulful. Someone mentions a fried fish joint up the road from where we put in, and we agree to meet there afterward. I ask Marshall if he has any parting words for us. He pulls his paddle out of the water and thinks. "We need people your age to take this job over and fight like crazy," he says. "People in my generation have fucked this up pretty badly. You know, when the worst hits, I'll be in that Great Duck Pond in the Sky. But you folks will have to figure this out. And the next decade will be critical—for the state, the country, and for the world."

I drove down to the town of Dulac, in southern Terrebonne Parish, Louisiana, on a brisk January morning. More than four months had passed since Hurricane Ida had raked the coast, but the signs of destruction were everywhere: crumpled houses, uprooted elms, a boarded-up elementary school. On my left, I was alarmed to see a house-size orange blaze leap from an open field, only to discover that this was one man's way of disposing of a damaged outbuilding of some kind. He looked on from a safe distance, shielding his face with a patterned tube scarf.

Cradled in the bayous, a stone's throw from a busy launch site called "Shrimpers' Row," Dulac is the kind of low-lying village that is most at risk from coastal erosion, the kind of place you might think Louisiana's master plan would be designed to protect. As of now, that proposition is in dispute. I think highly of the work that's gone into the master plan. I've praised it for being flexible, forward-looking, and fit for scale. But is it fair? We examined fairness in the context of climate resilience and caste in chapter 4. Sharon Lavigne, an environmental justice advocate, helped me understand what was personally at stake for residents in an overburdened region like Cancer Alley. I decided to go to Dulac, about forty miles south of Lavigne's home in St. James, to meet another climate justice advocate whose work has gained national and international attention: Shirell Parfait-Dardar, traditional chief of the Grand Caillou/Dulac Band of the Biloxi-Chitimacha-Choctaw Tribe (figure 6.2). She and other leaders of Native tribes located in this area have raised questions about the process and goals

6.2 Chief Shirell Parfait-Dardar at her home in Dulac, Louisiana.

(Photo by Rob Verchick)

behind some of the master plan's projects, including the previously mentioned Mid-Barataria and Mid-Breton Diversions. Parfait-Dardar had offered to tell me about the tribal residents on the coast and the concerns they have.

I first met Chief Parfait-Dardar about three years ago at a conference on coastal restoration. Most recently, we have worked together as members of the state task force I mentioned earlier, the one aimed at eliminating the state's net greenhouse gas emissions. Before our January visit, the chief had warned me her house would be under construction on account of the storm. On my arrival, I was greeted in the front yard by several men clustered around a table saw, eating sandwiches. Each wore a straw flat-brimmed hat, dark trousers, and suspenders. They were from Pennsylvania, one man said, all roofers, from the Mennonite Disaster Service, a volunteer network of churches that helps people recover from floods and storms. "There is a lot of need here," he said.

I knocked on the front door and the chief quickly invited me in. She suggested we sit at the kitchen table, pushing aside several neat stacks of receipts and insurance paperwork. A dressmaker, an activist, and a mother of four, Parfait-Dardar has a heart of gold and a backbone forged of tungsten steel. She was recently featured in a *National Geographic* documentary series hosted by Gal Gadot.[40] The episode's promotional description introduces its subject this way: "Meet Chief Shirell, the first female chief of her southern Louisiana tribe. Her ancestors have lived on the land for hundreds of years, and they are now expected to be among the United States' first climate refugees. Chief Shirell can't stop the storms, but she is impacting the legacy of her ancestors, the lives of her community today and the future of her people."[41]

"One 'Wonder Woman' meets another," I said.

"They did a really good job," she told me. "They had their crew down here for a week, and they understand what we're facing. And they promised they would be back."

For context, the Grand Caillou/Dulac Band has about a thousand registered citizens, who have historically lived in and around the ancestral village of Grand Caillou/Dulac in southern Terrebonne Parish, Louisiana. The Grand Caillou/Dulac are not formally recognized by the federal government. In fact, none of the coastal tribes are, a point that takes on more resonance in chapter 11. That means the tribe has no government-to-government relationship with Washington and is not eligible for the many federal grants and services specifically designed for Native communities. In contrast to many federally recognized tribes, the Grand Caillou/Dulac have no governing authority over any of their traditional lands, much of which is owned either by the state of Louisiana or by private entities.

Parfait-Dardar told me the Chitimacha People, from whom she is descended, have lived on the coastal plain "since time immemorial." In 1830, the U.S. Congress passed the Indian Removal Act, which led to the violent displacement of many southeastern tribes by way of the "Trail of Tears." To avoid relocation, members of the Biloxi and Choctaw tribes fled west to the Louisiana bayous.[42] "We amalgamated and basically became one people," Parfait-Dardar explained. "And we're here today." Like other tribal communities in southern Louisiana, the Grand Caillou/Dulac have traditionally lived on trapping, fishing, and farming.

The last hundred years of coastal destruction, combined with sea level rise, is seriously threatening the tribe's traditional practices. The master plan's projects, she said, are not focused enough on protecting the ecosystems they rely on. The Army Corps of Engineers, which collaborates with the CPRA, employs a sophisticated "science-based" system of "objective tools" to select the coastal areas most deserving attention. But this utilitarian focus steers the most ambitious projects toward the most populated areas. Of top concern are the wetlands that protect the New Orleans levee system from frontal assault. The Corps has told tribal leaders on the coast that the protection they seek is "too expensive and not sustainable."[43] Craig Colton, a geographer at Louisiana State University who has studied the state's restoration efforts extensively, shares many of the tribes' concerns and finds that the master plan does far too little to protect historically marginalized communities.[44]

"Fishing is not what it used to be," Parfait-Dardar told me. "We're losing these areas where shrimp come in, you know. They used to lay their eggs and you would look at the hatchlings—it's the cutest little thing. At one time, you could make a good living by being a traditional harvester or shrimper. You can't do that today." So tribal members seek other kinds of work. "You see a lot of our people get into the field of oil and gas which is killing us. That's what's available."

In addition to livelihood, Parfait-Dardar is concerned for her people's cultural heritage. "Our people were mound builders. Those mounds are still here today, especially in Grand Caillou," she said, referring to the nearby town. "There's a complex of nine mounds. They were made of shell and mud. They're amazing to see. Of course, we don't currently own those, which is kind of weird. It's owned by some land developers, but we're in talks right now."

Grand Caillou is situated on a natural plain called the Lafourche Subdelta, one of several major lobes of the Mississippi River Delta. Eight hundred years ago, these mounds, which follow the banks of Bayou Grand Caillou, would have provided enough elevation to support a village with worship and burial sites, meeting space, and trading markets. Now, because of saltwater intrusion, their foundations are beginning to erode and they are in danger of being swallowed into the gulf. Hundreds of ancient mounds on the coast are threatened by subsidence and sea level rise. Archaeologists estimate that Louisiana tribes lose two ancient mounds or villages a year.[45]

"We understand that the Army Corps has big projects," Parfait-Dardar told me. "Of course, you need big projects. But they're not considering smaller projects that could be implemented now that are very cost effective and can create immediate results." For instance, she says the tribes have proposed "living shoreline" projects that would armor the waterfront with natural materials while providing habitat for wildlife.

Parfait-Dardar is also organizing and seeking funds for a cost-effective initiative to backfill canals that were abandoned by the oil and gas companies and that pull harmful saltwater into the freshwater marsh. "We're definitely focusing on the ones that are closest to the sacred sites and that have the best potential for success. It's another means of healing the damage that was caused and trying to allow Mother Nature to repair herself." I pointed out that since 1972, federal law has required companies that dredge the canals to backfill them once they are decommissioned. Why not sue and demand that they repair the damage? "No," she said, "it could take decades before anyone would be held accountable. We're not willing to gamble with that. It's best for us to pursue this on our own."

In addition to what the Army Corps is *not* doing, Parfait-Dardar is critical of some of the things they *are* doing. Consider the Mid-Barataria and Mid-Breton Diversions. In the long term, these freshwater diversions promise to build new lobes of marshland that may be critical to the survival of the Greater New Orleans area. In the short term, however, the diversions will upset the current balance of estuarine ecosystems by converting areas of brackish water to fresh.

The change would likely decimate large populations of oysters, shrimp, blue crab, and many species of fish. The response from the Army Corps is that over the years, healthier fisheries will sufflate elsewhere. Besides, the water salinity on which today's fishers rely is not "natural" anyway, just the remains of poor human engineering. Parfait-Dardar's reply: fifty years is a long time to wait for more fish.

"We're not saying that everything that they're doing is wrong," said the chief, of the Army Corps. "It's just missing parts. We're already fragile and overburdened as it is."

"But the Corps," I say, pushing back gently, "says there's no way to make everybody happy, that some people will have to give up something so that a larger number of people can benefit." I already regret how this comes out.

"Well, I understand the concept, but that's bull crap. They don't look at the big picture. Look, you're not just affecting people," she says, referring to the communities that have lived on these shores for more than a thousand years. "You are also affecting estuaries and commercial fishing. There are many other projects that can be done. But talking to them is like talking to a brick wall." I ask her what she means.

"Well, you tell them, and they smile, and then they thank you, of course. But they turn right around and keep doing what they're doing. And it's like, did you not hear anything? Thousands of people are trying to tell you what's going to happen if you move forward with these projects." Colton has documented many experiences like this. Water-management projects in the coastal zone have a history of devaluing public sentiment, he concludes. And although protecting cultural heritage is a stated goal of Louisiana's master plan, Colton finds little evidence that this goal carries much weight.

"They've already made their minds," the chief sighed. "These meetings are a way of checking the box."

———◆———

Louisiana's restoration plan relies not only on exquisite feats of engineering, but also on a radical rethinking of community, culture, and place. All the diversions and pumping and dumping are impressive. But even if it all works, the lives and livelihoods of thousands of people and many unique cultures will be disrupted. Some communities will have little choice but to move, a topic we will take up separately in chapter 9. What's learned here will influence coastal resilience projects throughout the United States and the rest of the world.

A main feature of the master plan is supposed to be its promise to "balance" a range of political objectives: to "provide flood protection, use natural processes, provide habitat for commercial and recreational activities, sustain our unique cultural heritage and support our working coast."[46] But the state is breaching that promise.

Journalist Nathaniel Rich, a New Orleans resident, gets to the root of the frustration in a piece he wrote for the *New York Times Magazine* in 2020. Referring to Louisiana's aim to balance all interests, he observes: "This was unimpeachable in theory, offering something for everyone. But the plan was silent on what to do when these objectives came into direct conflict. What happened when flood-protection measures threatened cultural heritage? Most maddening of all was the plan's emphasis on the future over the present."[47]

Rich underlines two types of conflicts, which we learned about in chapter 3, and that often accompany resilience efforts. The first is a clash of values. As a society, we want government programs that are evidence-based, cost-effective, and that deliver the most bang for the buck. In the first half of the last decade, water management projects led by the Army Corps seemed almost the opposite of that. Levees and dams were instead doled out like candy by some members of Congress seeking to reward local politicians, land speculators, or some other ally with a sugar tooth.

To inspire public confidence, the Army Corps submitted itself to its own straitjacket of sorts: a set of decision-making criteria that emphasizes economic utility above nearly any other value. One of those subordinated values is "justice," which I described in chapter 4 as meaning that the government has a special duty to address the interests of disadvantaged groups, most notably those bearing the scars of caste-based mistreatment. There are surely permissible ways for the U.S. Army Corps of Engineers and the state's coastal authority to balance the interests of efficiency and justice. But the deck is already stacked— by presumptive economic formulas, agency culture, and resources (the Corps employs many more economists than anthropologists, for instance).

The second conflict is about scale—not geographic or jurisdictional scale, which I think the master plan handles well, but *temporal* scale. How do you balance the needs of today's fishers against the needs of people fifty years from now, most of whom are not even born? This philosophical puzzle has put more than a few ethicists in therapy, and I don't have a solid answer. But I do think balancing interests across generations requires both a robust and expansive deliberative process *and* a commitment to provide sufficient resources to compensate, retrain, and, if need be, relocate, those in the current generation whose interests must give way. I asked in chapter 1 whether there should be any "losers" on the road to resilience. Here, I'm saying there will certainly be people whose interests are subordinated to others. Whether they should be called "losers" will depend on how well society recognizes and addresses their circumstances and needs.

Louisiana's moonshot on the bayou deserves high acclaim. This state, once again, is offering some of its best and brightest to the world. This time it's not drilling technology, or fishing prowess, or even jazz. It's a plan to restore—at least for a substantial period—one of the world's most imperiled and productive ecosystems. Still, it must be improved if it is to make good on its commitment to cultural heritage and the needs of all Louisianans.

Governance is tough. Chief Parfait-Dardar told me so. Parfait-Dardar was appointed chief of the Grand Caillou/Dulac when she was twenty-nine years old. Her uncle had been the tribe's previous leader, and he had decided to retire. "I was so scared," Parfait-Dardar remembered, "because I knew the shoes that I had to fill. I was the first female in a very long time to take on the task of chief. I was terrified, you know. I remember, I asked my uncle, 'What if I mess it all up?' And he said, 'Look, if you mess it all up, then the only thing you can do is fix it.' That's pretty good logic. You can't be afraid to fail. What you should fear is not trying."

If you mess it up, you fix it up. To the moon!

chapter 7
Lights Out

When I dream about Hurricane Katrina (and I still do), it always starts with the refrigerators. Kenmore, GE, Whirlpool, Frigidaire, Amana. Hundreds of thousands of these abandoned appliances stood duct-taped shut on the curbs and yards of homes throughout New Orleans. Many were spray-painted with whimsical or foreboding messages. "Funky. Not in a good way." "Free Beer and Maggots." "Smells like FEMA." "The Bowels of Hell await you within!!"

In my dream, there are no people, just endless lines of aluminum tombstones planted in muck, leaning this way and that. There's a hot breeze, and the whole neighborhood smells like low tide times ten. Dragonflies are buzzing everywhere. I think about going inside the house, but it's more than 100 degrees in there. There's no air conditioning: the power's been out for weeks.

Katrina refrigerators (it's a thing, Google it) are not the only or even the most dramatic example of the perils of power outages in extreme weather. But that meme is what haunts me, of course, because I lived it. In any case, the scene is less abstract than the loss of "10,000 fish representing more than 530 species" at the city's famous Aquarium of the Americas, and less torturous than the chaos at Memorial Hospital, where floodwaters marooned dozens of elderly patients on sweat-soaked bedsheets, without electricity or running water, until they were finally rescued or died.[1]

There are lighter moments, too. When an appliance-disposal worker was interviewed about the smelly job of emptying abandoned refrigerators of their oozing contents, he spoke a few words about wind direction and shrugged, "I'm a retired mortician, so it don't bother me much."[2]

From rancid food to emergency-room nightmares, communities take a punch when the lights go out. The nation's aging power grid leaves us more susceptible to such risks. The growing intensity of floods and storms on account of climate change make things even worse. But Katrina was not the only, or even the loudest siren blow. Since that time, storms have smashed or drowned parts of the grid in the Northeast (Hurricane Sandy), Houston (Harvey), the Southeast (Irma), and, most tragically, in Puerto Rico where in 2018 Hurricane Maria knocked out 80 percent of the island's electricity network, causing the largest blackout in U.S. history. In the same decade, wildfires in Northern California focused attention on the neglected and antiquated system operated by Pacific Gas & Electric, which is believed to have ignited several devastating blazes over two years. Killing at least eighty-four people and destroying billions of dollars in property, the fires also required cycles of deliberate blackouts to protect parched forests from even more incendiary failures. In this way, the grid became both a cause and a victim of violent disaster.[3]

We need a power grid that is more resilient—one that has the capacity to cope with the kinds of *low-frequency, high-impact events* like those described above. In this chapter, I hope to show how we might get there. I'll explain what the grid is, describe some of the grid's most serious vulnerabilities—emphasizing the link to climate disruption and the outsize risks faced by marginalized groups—and propose some first steps toward making the grid more resilient and more productive for all.

It has been called "the world's biggest machine." But the U.S. power grid is more accurately understood as an elaborate and tortuous enterprise consisting of about 7,000 power plants, 55,000 substations, and six million miles of wire. It was not intended to be nimble and spry, but instead, static and stable. Its job was to produce electricity that was reliable, geographically accessible, and affordable.[4] Those goals, which were always in conflict, have now been complicated by two additional needs—that power be *clean* and that it be *resilient* in the face of more modern catastrophic risks, including storms, floods, and drought.

One way to conceptualize the power grid is to think of it as an enormous bathtub, filled to the brim with water, with a goal of keeping it that way. The water represents all the electricity available on a national or regional transmission *network* (one for the western half of the country, one for the eastern half, and one just for Texas). There's a big drain at the bottom of the tub, and it's always open, forever funneling energy out of the system. That drain is us, *users* of all shapes and sizes—households, hospitals, Amazon server farms, and more. Without a full tub, consumers won't get their shares on time, so we need bucket brigades lined up around the tub constantly pouring more and more energy into the bath. (If you're imagining Mickey Mouse as the "Sorcerer's Apprentice," you've some idea of the complications involved.) Maintaining the equilibrium—the "waterline" in the tub—is extremely important for the process to work smoothly and reliably. If one bucket brigade trips up or collapses, that team immediately must be hauled out and replaced with another one. Those brigades, who represent facilities like power plants, hydroelectric dams, and wind farms, are the *generators*.

Since the day Thomas Edison's Pearl Street station began lighting lamps in lower Manhattan, the relationship between *generators*, *networks*, and *users* has been remarkably stable. There was a linear progression from "high-voltage" activities to "low-voltage" ones. At the high-voltage level were the generating plants and the interstate transmission lines. The generators sold electricity to the operators of the high-voltage lines, which carried electricity miles away to where it was needed. (In the past it was common for utilities to own both the plants and the lines—in effect, selling power to themselves—but today, because of regulatory changes, many utilities now own only the lines.)

The low-voltage level activities made up a "distribution" system in which high-voltage current was "stepped down" at substations and then delivered at lower voltage to a local network where it was stepped down once more before entering the customer's home or workplace. Those step-downs are what keep your toaster from exploding at the breakfast table. Even if you didn't know before now what a substation does, you've seen them. They are the clusters of powerlines, breaker boxes, and curlicued transformers sitting at the edge of neighborhoods or small towns. Usually, they're encircled by a wall or a chain-link fence. My middle son, when he was younger, said one looked like a cage of giant hair curlers. As for the local distribution network, if it's underground, you may never see it. If above ground, it is entwined with the cable and telephone wires draped across the neighborhood utility poles. (The second stepping down of voltage I mentioned occurs in those overhead cylindrical transformers, so attractive to squirrels and crows.)

But now things are getting more complicated. Today, localized technologies like rooftop solar generation allow users to also act as generators in distributed energy systems. Digital systems embedded in transmission networks can now control how much power commercial users request at certain times or how much power generators will produce, which gives the network some characteristics of both the user and the generator.

Power grids have always been vulnerable to natural disaster. Now, that vulnerability is backlit by climate change. A heating planet can make floods higher, storms fiercer, and water flows used for cooling and power generation less reliable. There are other risks, too, which we are only beginning to study in comprehensive ways. Experts seem particularly worried about the potentially catastrophic effects of physical sabotage or a cyberattack. Geomagnetic storms induced by the solar winds are also a surprisingly urgent concern. (For our purposes, we'll save deliberate attacks and "space weather" for another day.)

To survey the risk, we can start with generation. When Hurricane Sandy walloped New York City, one-third of its electric generating capacity was knocked out of service. That's not a surprise when you consider that the city's gas-fired power plants are all perched on the waterfront, some more than seventy years old. Because of that and transmission failures (37 percent of the city's substations are prone to flooding), two million New Yorkers lost power by the time the storm had passed.

Low-lying power plants serve not only New Orleans and New York, however, but also Seattle, San Francisco, Los Angeles, Houston, and more. According to a report commissioned by the Union of Concerned Scientists, "some 100 electric facilities in the contiguous United States, including power plants and substations, are sited within four feet of local high tide. And as sea levels continue to rise, the risks to these facilities from storm surge and floods will also increase."[5]

Why are so many power plants near the coast? The question reminds me of the time, many years ago, that I visited a nuclear power plant in Forsmark, Sweden, on the Baltic Sea. Having inspected the plant's Lego-like containment structures, rambled through fluorescent-lit labs, and even descended about two-thirds of a mile (1 km) underground to view the plant's radioactive waste repository, we legal academics were all duly impressed. At the end, our host, a soft-spoken Swedish engineer, concluded the tour with an observation both modest and wry. "My friends," he said, "what we have witnessed here is just a very complicated way to boil water."

This is actually a very important point. Most of the ways we generate electricity—from burning carbon to splitting atoms—are just complicated ways to boil water. The boiling water makes steam, and the steam rotates turbines to produce electricity. However you do this, the process requires lots and lots of water, not just to have something to boil but also to *cool the steam*. This way, it can be condensed back into a liquid and boiled again. That's why Sweden's Forsmark plant is located on the Baltic Sea, an endless source of liquid coolant. And that's why so many U.S. power plants of all kinds recline on low-lying coasts. The irony: the seaside location that makes power generation so practical is exactly the characteristic that climate disruption will make so vulnerable.

Surge and storm are not the only big challenges for steam-generating power plants (also called thermal power plants). Energy experts worry about heat and water scarcity. Between 2006 and 2013, researchers have found dozens of instances in which power plants were compromised either because there was not enough water or because the water was too warm. Most failures occurred in the summer when households needed electricity most. For instance, in the summer of 2012, water levels at Iowa's Cedar River fell so low that operators at the Duane Arnold nuclear plant had to dredge the river to corral enough water. Summer droughts in 2007 and 2008 put Wyoming's Laramie River Station, a coal-fired power plant, at risk of shutting down for lack of water. Officials had to divert water from farmers to avoid rolling blackouts. In August 2012, the Millstone Nuclear Power Station in Connecticut was required to power down a reactor because water pumped from the Long Island Sound was too warm to cool the steam, which resulted in several million dollars of loss. Three times in the last decade, the Tennessee Valley Authority was forced to shut down Alabama's Brown's Ferry nuclear plant because the used cooling water was too hot to discharge in the already overheated river without threatening ecological damage. Again, many millions of dollars were lost. Overall, experts say about a quarter of existing thermal power plants are in counties that by 2030 will be prone to high-to-moderate water supply problems.[6]

All this is enough to make you wonder if boiling liquid to spin turbines might be a bit overrated. Renewable technologies like hydroelectric power, solar energy, and wind power surely have big advantages in the climate crisis. We will get to those. However, climate-change impacts will also test the resilience of some renewable technologies. Hydroelectric plants, which are powered by waters rushing through penstocks and spinning turbines, are very sensitive to changes in rain patterns and evaporation rates. Consider Lake Mead, which funnels water from the Colorado River through seventeen enormous generators at Hoover Dam. Were the lake to completely dry up (a plausible scenario), more than a million people would need to get their power from somewhere else. An important advantage of renewable sources like wind turbines and the photovoltaic cells used in solar panels is that they don't rely on large volumes of water to make steam or cool fluid. (Some solar technologies, including some that use mirrors to focus heat to generate steam, do use large amounts of water.)

You'll notice that we are only talking about the *electricity* part of the energy supply chain, not fuel production, storage, and transportation. Today almost 60 percent of our electricity comes from two fossil fuels—natural gas and coal. Extracting and transporting those things gets harder as the climate unwinds.

Perhaps you remember Winter Storm Uri, the cold snap that struck Texas in February 2021.[7] Statewide energy disruptions followed, leaving as many as four million people without electricity or heating fuel in icy temperatures. Water mains burst, and water treatment plants failed. All told, more than a hundred people died. The main culprit? Failed power production—specifically from natural gas, which fed two-thirds of the state's generators. When temperatures plunged, pumps at the wellheads and pipelines froze. Gas output shrank by 45 percent and power plants across the prairie hungered for fuel. Some gas-fired power generators—with their own pumps and pipes to deal with—were also immobilized.

Without evidence, some officials and conservative news outlets blamed the mess on "ugly wind generators."[8] In fact, wind—which makes up only 7 percent of the state's winter capacity—performed as well or better than expected. At any rate, almost none of the electricity-producing infrastructure, from pumps to pipelines to plants to wind turbines, had been equipped with adequate insulation or heating elements. Though aware of such vulnerabilities, state regulators have long refused to require "winterization" for gas and power networks for reasons of cost. The Texas plan is to go naked.

But it's not just inland Texas. Offshore, the Gulf of Mexico is home to nearly four thousand oil and gas platforms. These structures, many of which were never designed for rising seas, are vulnerable to extreme wind and surge. (Katrina alone wrecked a hundred of them in 2005.) Half of the nation's natural gas processing and oil refining is, of course, located on the surrounding coast and vulnerable in similar ways.

Once the electricity is generated, all that energy jitterbugs through miles of high-voltage transmission cable—each atom in the line snatching an electron and flinging off another—until a current of 120 volts funnels into your home. When it gets too hot, those uninsulated, high-voltage transmission cables stretching miles across the American landscape can sag. My grandfather knew that. Following the Great Depression, he left his home in Oklahoma, where his brothers were toiling among the state's oil derricks, for the dream of attending art school in Los Angeles. He would do that, but not before completing a series of labors along the way, one of which involved a summer hanging powerlines in the Arizona desert. I loved the stories he told me as a kid and fell in love with the Glen Campbell hit "Wichita Lineman."[9] I can still see the shadows of arced powerlines that I imagined stretching across the vacant sands my grandfather knew. But I never understood the process that caused powerlines to sag or why it mattered until I began researching climate change.

It's all about physics. The cables—which are made either entirely of aluminum, or of steel surrounded by aluminum—lose their rigidity as they heat up. The heat is a product of the high electrical current and the ambient temperature. When the metal simmers long enough, the cables go limp. It's a common enough occurrence, but not without its downsides. To begin with, hot cables are less efficient. The metal's electrical resistance increases with the heat, and this makes it harder for the current to push through the lines. In perfect conditions, transmission lines already lose about 7 percent of the energy they transmit, but this loss climbs quickly with the temperature. Think about the summer temperature spikes climate change is already causing. Add that to a soaring energy demand necessitated by more air conditioning, and you begin to see the problem. In this punishing environment, the more the workhorse works, the more tired and less efficient it becomes.

More harrowing is the possibility of a "flashover," in which a cable's high-voltage electricity arcs to a nearby tree or other grounded object, shutting down the system and causing fires and explosions. On one hot August afternoon in 2003, a sagging power line in northern Ohio grazed an overgrown tree and did just that. An alarm that should have notified network operators didn't. Three more flashovers occurred shortly thereafter, and the additional current was forced onto a grid too weak to accept it. In a matter of minutes, power plants began failing throughout the Midwest, and in less than two hours the entire system had collapsed. In Canada and eight northeastern states, fifty million people lost power for days, for some even weeks. Experts attribute eleven deaths to the event and economic losses of six billion dollars.[10]

But it's not just about sag. Storms and wildfire also pose serious threats to the power grid. Since 2000, weather-related disruptions like these have been getting worse. In 2012, to name one example, an epic band of thunderstorms, known as a derecho, killed the lights for nearly four million residents across the Midwest and Mid-Atlantic coast. Uncontrolled blazes will of course incinerate transmission poles or cause trees to knock them down. But the greatest risk comes from all the smoke and ash, which "ionizes" the surrounding air, giving it an electric charge. Ionized air creates an electrical pathway allowing current to jump from transmission lines and crash the network. In just this way, the 2011 Las Conchas wildfire in New Mexico nearly debilitated two high-voltage transmission lines carrying power to 400,000 customers.

Everyone relies on plentiful, reliable, affordable power. But, when things go wrong, it is the poor and otherwise disadvantaged who suffer most. Studies consistently show that in disasters of all kinds, poor people and people of color are more likely to suffer property damage, bodily injury, and death. Power outages discriminate in this way by stealing the most necessary services from the populations most dependent on them. Think of all the things we lose when we lose power: light, temperature control, drinking water, sewage disposal, food storage, medicine storage, transportation, communication, hospital care—even, as one study notes, the ability to maintain mental health. Post-traumatic stress disorder, to name one condition, is common in many kinds of disaster situations. We all need these things, but some populations are more dependent and less able to find substitutes when familiar supports are cut off.

Much of the literature on power outages focuses on hospital services. The tribulations at Memorial Hospital in New Orleans, mentioned earlier, make the point. Researchers have cataloged an array of impacts in hospitals related to blackouts, including disruptions to direct clinical care, cleaning and sterilizing processes, patient record keeping, transportation systems, and emergency communications. At the same time, when power goes out in a city, the demand for hospital service skyrockets. These problems obviously affect certain populations more: the disabled, people with chronic illnesses, the very young, and older adults.

Other important disparities are related to heat extremes in crowded cities. Despite the impression you get from all those wind-battered weather announcers on the local news, heat waves are a much more lethal weather-related killer. Every year in the United States, hundreds die from such events. In July 1995, a searing heat wave in Chicago killed 739 people, roughly seven times as many as died in Superstorm Sandy. The impact can be particularly severe in dense urban areas, where asphalt and concrete surfaces absorb heat and tree cover is sparse. Because of this phenomenon, known as the "heat island" effect, the annual average air temperature of a city with one million people or more can be 1.8 degrees to 5.4 degrees F warmer than its surroundings. In a survey of more than 304 metropolitan areas, researchers from the University of California at Berkeley found that Black people were 52 percent more likely than white people to live in such densely packed neighborhoods, while Asians were 32 percent more likely, and Latinos 21 percent more likely to live in such areas.[11]

Blackouts during a heat wave are especially dangerous for anyone living in a heat island, particularly the elderly and the disabled. Imagine being trapped in a sweltering high-rise building with an inoperable elevator and no tap water because the pumping equipment has failed. Now, imagine the phone network is down, and you can't call for help. Given future trends in population growth and urbanization, these problems are likely to increase.

Clearly, we need a power grid that is more resilient—one that has the capacity to cope with the kinds of *low-frequency, high-impact events* like those described above. Resilience differs from what electric companies call "reliability." Reliability is about smaller, more common disruptions like a falling tree or failed equipment. While utilities and regulators do focus on some elements of resilience, the efforts are piecemeal. For example, while federal regulators have recently devoted more attention to long-term fuel disruptions, cybersecurity, and geomagnetic disturbances, it has all but ignored the threat of climate disruption. And resilience is rarely an explicit concern in formal utility planning.

The way to make the power grid more resilient is to pay more attention to its durability and flexibility. We can divide that work into three categories: *hardening* the grid, *smartening* the grid, and *greening* the grid. Hardening refers to the mostly physical fixes that protect equipment from weather-related damage.

Smartening the grid means using digital technology to make the delivery of electricity more responsive, efficient, and flexible. Smart-grid technology also helps lower user demand, adding again to network resilience. (A grid with lower burden is a grid that fails less.) Greening the grid is my term for integrating more renewable energy sources into new and existing electricity networks. Some kinds of power generation, like wind and solar, turn out to be more physically resilient in storm conditions than their carbon-burning cousins. Because renewables reduce greenhouse gas emissions, they also mitigate the future impacts of climate change.

There are many ways to physically protect important parts of the grid from extreme events. Such work might include building berms and levees, restoring protective marshes, elevating substations, burying distribution lines, or even trimming trees near high-voltage cables. Installing diesel backup generators in buildings and homes fits in this category, too. In Northern California, Pacific Gas & Electric is working with state and federal agencies to restore bay habitat to dampen tides near shoreline facilities. The company uses new "dry cooling" technology in two of its natural-gas plants to minimize water needs. Florida Power and Light now actively monitors climate-change-based flood risk at its Turkey Point nuclear facility on Florida's Biscayne Bay. The current power plant is elevated eighteen feet above sea level, and all equipment is shielded against waves up to twenty-two feet. Plans to build two new reactors on the site will consider long-term projections on sea level rise, hurricane surge, riverine floods, tsunami hazard, and dam failure.

In December 2012, President Obama established the Hurricane Sandy Rebuilding Task Force to provide the coordination necessary for the rebuilding of public housing, transportation, and utilities. The resulting strategy includes sixty-nine recommendations, addressing everything from promoting a more resilient electricity grid, to hardening communications infrastructure, to assisting state and local governments in accessing capital.[12]

But there have been setbacks. In the summer of 2017, President Trump launched a campaign to improve the nation's infrastructure with an executive order that rescinded an Obama-era standard requiring constructions projects that rely on federal money to factor in future conditions—including climate-change impacts—when building in floodplains.[13] Thus, when Texas and Florida began the process of rebuilding damaged communities after back-to-back hurricanes that same year, there were no rules in place to require engineers to build back better with climate change in mind. (Fortunately, President Biden reinstated the Obama-era flood standard on his first day in office.)

On the Gulf Coast, power outages are a common concern in country towns and urban neighborhoods. I write my checks to Entergy, a private power company based in New Orleans that generates and distributes electricity across several southeastern states. Its network along the Gulf Coast is famously vulnerable to land subsidence, sea level rise, and hurricanes. In 2007, for the first time, Entergy began forecasting the future damage to the coast under current trends. The results are terrifying. The company estimates that by 2050 the *annual* damage to assets on the coast (including homes, businesses, infrastructure) will total $40 billion. Of that, the oil and gas network will suffer $7 billion of damage each year, and the power grid will take a billion-dollar hit.[14]

Here I recall a conversation from 2016 that I had with Charles Rice, who was at the time the CEO of Entergy-New Orleans, a subsidiary company. He told me that climate became a priority for Entergy after Hurricane Katrina. Management saw that neither Entergy nor its customers could thrive in the Gulf of Mexico unless private companies and the government started planning better, restoring the coast, and hardening infrastructure. "Everybody has to stack cans," he said. He listed several initiatives that Entergy was engaged in, from armoring a plant just east of New Orleans to replacing wooden utility poles with steel ones. The utility was even trying to reforest swamps by spraying mangrove seeds from crop dusters. I was skeptical that that was enough and told him so. Rice's answer still sticks with me: "What can our customers afford?"

In a city where a quarter of the residents live in poverty, a customer's ability to pay is obviously a top concern. Recall that cost was a reason given by Texas regulators for refusing to require gas and power producers to insulate their equipment. The problem of energy affordability, often referred to as "energy insecurity" is real. According to federal data, nearly a third of U.S. households had trouble paying their energy bills in 2015. One in five gave up necessities like food or medicine to keep the lights on. Black people, Latinos, and Indigenous people are much more likely to suffer these hardships than are white people.[15] All this is shameful and unacceptable. And it goes *way* beyond preparing for climate change. Which is to say we should not accept the injustice of energy insecurity as a "given" that climate resilience must work around. Indeed, a vulnerable grid is energy insecurity in just another form.

Of the major utilities doing the most to protect its assets from climate disruption is Consolidated Edison (Con Ed) which provides electricity to New York City and Westchester County. The company, which traces its roots to Thomas Edison's Pearl Street power station in lower Manhattan, has a long history of battling extreme weather. In the 1880s, when city streets were canopied in a spaghetti of power and telegraph lines, Edison insisted that currents running from *his* generators travel through wires *underground*. The decision, motivated by safety concerns, paid off. In the famous blizzard of 1888, when most of Gotham went dark and immobile, Pearl Street customers carried on beneath soft flutters of lamplight.[16]

But wait, you say, there were bad times too—like those two million New Yorkers who lost power in the wake of Hurricane Sandy. Remember all those flooded gas plants and substations we saw in the beginning of this chapter? Well, Sandy was the alarm bell that Con Ed is now heeding. After that storm, the utility launched state-of-the-art measures to plan for and protect the power grid from future climate damage. We need to look at the utility's reaction here for three reasons: first, to identify what catalyzed Con Ed's race toward resilience; and second, to see what state-of-the-art resilience even looks like. Somewhere in there, we'll tackle the puzzle I posed in chapter 4 about protecting substations without knowing how high the sea is going to rise. *Trigger warning*: this story is driven by young activist lawyers.

Let's back up.

On the night of October 29, 2012, Hurricane Sandy plowed into Manhattan with a wind and surge that demolished a large chunk of the city's power grid, including half of Con Ed's high-voltage power lines and five transmission substations serving more than a million customers. (Remember the caged hair curlers?) At Con Ed's East River substation in lower Manhattan (to pick one example) the water unexpectedly topped thirteen feet—just two feet above what the station's critical equipment had been designed to withstand. That night, the station's manager described a pale-blue arc of electricity leaping from a circuit breaker into the sky just minutes before lower Manhattan went dark. "I was pretty much speechless," he told a reporter, standing outside the plant years later. "If I wasn't here to see it, I wouldn't believe it."[17]

It had taken the utility more than a week to restore power to the city, and there loomed a mountain of repairs on the horizon. Con Ed quickly went to work, proposing a billion-dollar construction project to rebuild and fortify the city's network against another Sandy-like storm. There was just one problem: no one was thinking about how an unspooling climate would make things worse *in the future*. Sandy, for all its sound and fury, was already an artifact of the past. It wasn't that the engineers didn't know or care about future climate disruption. By then there was good evidence that global warming would continue to super-size Atlantic storms. But it was hard to know how bad it would get. Plus, whatever the utility did was going to cost more money. Where would that come from? And who would decide what's "good enough," anyway?

To conceptualize the grid, we've been painting mental pictures from bathtubs to jitterbugging. Now let's imagine a trapdoor, perhaps in the flood-buckled floor of that East River substation. I open it. You follow me down into the blank space below. Here I will tell you a secret that underlies how all retail electricity in the United States is produced—the truth about how an electric company's *profits* are made. After this, you will grasp why hardening the grid is so hard and what we can do to make it easier. That same knowledge will suggest ways to smarten the grid and make it greener too. What's more, at your next cocktail party I promise you will know more than anyone else about what keeps the chandeliers lit and the icemaker churning.

Down the hatch.

If you are like me before I began writing about energy law, you might think utilities make their money by selling electricity. They don't. The law, in fact, prohibits that. Instead, all or most of a utility's profit is based on building stuff—substations, transmission towers, nuclear power plants, and so on. Why? History.

Once Edison's Pearl Street Station breached the gloom of New York office space in 1882, the utility business erupted into a frenzy of competition that lasted until the early 1920s. That is when the industry's small fry were suddenly swallowed up by a pod of cigar-chomping monopolists, including the bespectacled (and indicted) Samuel Insull, an early associate of Thomas Edison. By 1925, it is said, three-quarters of the U.S. electricity business lay in the hands of eight holding companies. A decade later, under threat by trust-busting politicians, the industry brokered an agreement by which utilities would be given regional monopolies in exchange for submitting to government reliability standards and price controls. The deal, known as the "utility consensus," continues to this day.

Almost everywhere in the United States, price controls operate under a principle called "cost-of-service." Being state-controlled monopolies, utilities are not allowed to profit from their main product. Instead, they sell power to ratepayers without any markup. To induce private investors to put up capital to improve and expand infrastructure, regulators offer them a predictable markup, or "rate of return," on the utility's capital investments. Under this system, all "prudent" capital investments are assured a reasonable return. Deciding what investments are "prudent" and what rates are "reasonable" is a job that falls to a state's public utility commission. That determination occurs during public hearings in which representatives of the utilities and members of the public get involved. These hearings make up what is called a "rate case."

The reason rate cases are so important is that they determine what kinds of improvements a utility will want to invest in. If the utility commission rewards building waterproof control rooms, companies will do that. If it rewards expanding the power networks into dangerous flood zones, companies will do *that*. If the rate of return *minimizes* the money that can be made from greening the grid, electric companies will take a rain check. All the while, the power companies advocating for new construction know more about their proposed projects than either the commission or outside participants. Plus, relations between utilities and commissions are notoriously cozy. Anyone can learn about rate cases by visiting their utility commission's website and snooping around. They're not supposed to be secret, but most people have no idea rate cases exist. Now, happily, you do.

Up through the trapdoor and back into the fresh air!

When we left off, Con Ed was planning to spend a billion dollars to rebuild and fortify the grid against another Sandy-size storm. But, as we just learned, Con Ed first needed permission from New York's utility commission to raise its rates so that it could obtain a reasonable return on investment. At the subsequent hearings, representatives for Con Ed made their case and a few consumer groups also appeared. But, still, no one was talking about climate disruption.

Enter a newly licensed, activist lawyer named A. R. Siders. At the time, Siders was in her twenties. She had a law degree from Harvard and had recently completed a fellowship with the U.S. Navy, where she had helped prepare military facilities for climate change. Siders had just moved to New York to work at Columbia University's Climate Law Center (now called the Sabin Center for Climate Change Law) when Hurricane Sandy upended the city. As the state began to recover, Siders's boss, a Columbia law professor named Michael Gerrard, suggested the idea of intervening on behalf of the center in the Con Ed rate case. No one on at the hearings was talking about climate change. Maybe they should have been.

Siders agreed and went to work helping to build an alliance of environmental and green energy advocates. The alliance, which came to include Pace Law School's Climate and Energy Center in New York, the Hudson Riverkeepers, and even the Municipal Art Society of New York, not only participated in the Con Ed proceedings but further petitioned the utility commission to use its regulatory authority to insist that *all* utility companies in the Empire State evaluate and plan for climate damage. "Infrastructure that has historically been safe from extreme weather events cannot be assumed to be safe from future events," wrote Siders in the group's letter of petition. Preparatory action was needed "to ensure the reliable provision of vital service to New York citizens."[18] Note how Siders draws a line from the "reliable provision of vital service"—a phrase pulled from the "utility consensus" of the New Deal era—to our modern climate crisis. That's more than a lawyer's flourish. Siders was reminding the commission that while circumstances may change, the sacred bargain between public utility and citizen sovereign must always endure.

When I graduated from law school years ago, there was nothing I learned in my three years of study that would have acquainted me with utility rate cases or administrative petitions. What did Siders know about such proceedings? "Not a thing," she told me. "I'd finished my law degree. I'd worked for the Navy for two years. This was totally new to me." But she and her colleagues, who included a posse of more young activist lawyers and law students, pushed forward, researching statutes, interviewing climate scientists, and haggling with Con Ed representatives. Everything they learned suggested the city needed to be ready for storms even bigger than Sandy.

But Con Ed's experts were saying something different. "Oh, we're going to build flood walls, and we're going to build them to *Sandy flood levels,*" said Siders, channeling a company engineer she talked to. "But anyone who works in climate change would look at that and say, 'No, no, no. The answer needs to be Sandy-*plus.*'"

But Sandy plus what?

"That's a hard question," she said. And here we stared at each other for a moment or two. "One foot? Two feet? Either of those is reasonable," she said. "But they have different costs." Then: "We didn't have a clear answer."

People disagree about risk depending on where they sit. "Con Ed may have a very different risk tolerance as a company than that of the people they're serving, or of the city government," Siders told me. "How much risk are we willing to bear? How much safety are we willing to pay for? We don't talk about these things very well."

Recall that before Siders and her colleagues came along, the participating community groups were mainly focused on their pocketbooks. "A number of them opposed the stuff we were saying about climate change because they were concerned it would raise the cost. I mean, think about anyone on a fixed income. That's a huge challenge."

I think now of Entergy's Charles Rice: "What can our customers afford?" Sometimes life seems to me one delicately branched decision tree rooted in a sandbank of risk. Do you choose lower rates and lose power during a freak storm someday in the future? Or do you default on your utility bill and have your air conditioning shut off tomorrow? Sure, we all have to stack cans. But how high?

Would you believe *three feet*?

You see, after months of negotiations, Con Ed eventually agreed to revise its design to ensure that facilities and equipment could withstand a "one-hundred-year flood," as defined by FEMA, *plus* an extra three feet.[19] In addition, Con Ed pledged to work with the state attorney general, the mayor's office, and a team of environmental groups, to develop a long-term, scientifically informed, climate resilience plan. The commission, in exchange, granted Con Ed a 9.2 percent return on its billion-dollar investment. To protect customers from a spike in energy bills, the commission set rates in way that provided households a credit in the first two years, which could be used to offset rate increases after that.

By 2018, Con Ed had completed most of these fortifications, which included the complete reinforcement of twenty substations. At the East River station, the control room was elevated several feet and perimeter walls were hardened. A 345-kilovolt, "octopus-like" pass breaker now allows transformers to be shut down separately in case of emergency. "Where water rushed past steel doors," according to one observer, there are now "fast-deploying Kevlar sheets that look like large, industrial-strength shower curtains."[20] High-volume flood pumps guard vital equipment in case those shower curtains fail. To temporarily waterproof the station's exquisite control panels, operators are issued pre-sized, military-grade shrink wrap.

The important thing to understand about what became known as the "FEMA plus three" standard is that it was basically off the cuff. The city needed a rebuilt grid pronto. Fine-tuned data was unavailable at the time. So, the utility took the FEMA flood measure, added a two-foot margin of safety (what engineers call "freeboard"), then slapped another foot on top to account for a modest rise in sea level. As we say in New Orleans, that was "close enough for jazz."

No one knew if "FEMA plus three" was enough; in fact, it probably wouldn't be. But the exercise *bought time* until a robust climate assessment could be performed—the more brooding, symphonic version of resilience, if I may continue that metaphor. When the robust assessment was complete, the remainder of Con Ed's development plans would be revised accordingly.

The climate study was completed in 2019 and heralded by legal experts as "the most robust climate resilience planning effort undertaken by any electric utility to date."[21] The study analyzed projected changes in temperature, humidity, precipitation, and sea level rise from 2020 to 2080. Con Ed consulted with scientists at Columbia University's Lamont-Doherty Earth Observatory and other organizations to downscale projections for the New York area. To account for uncertainty, the study relied on carefully considered scenarios that model how variables like temperature, precipitation, and sea level rise might change.

The study, which has now been developed into a planning model, deliberately leans away from rosy pictures. Drawing from the upper range of plausible outcomes, the model assumes that by 2080 summer days in Central Park will average 85 degrees F, ten degrees hotter than today's mean. By the same year (using a midrange number), the model forecasts the sea to be twenty-eight inches higher. This is considerably more than the one-foot allowance adopted in the original "FEMA plus three" standard. Indeed, the design standard for Con Ed's East Hudson substation—Kevlar curtains and all—could be exceeded in less than twenty years. By then, East Hudson may either need to be refortified or abandoned. New facilities intended to last through the century are already being built to withstand these more extreme projections.[22]

Critically, the climate information relied upon will be continually updated, leading to continual revisions of building standards and development maps. Because of information gaps, Con Ed's model does not yet consider underground temperatures (which affect buried lines) or wind behavior. But as more science comes in, those variables will be added.

In approving the settlement that catalyzed Con Ed's planning, the utility commission called on the state's *other* utilities to hop on the resilience train too. "We expect the utilities to consult the most current data to evaluate the climate impacts," wrote the commission, "to integrate these considerations into their system planning and construction forecasts and budgets."[23] While encouraging, these words fall short of *requiring* all New York–based utilities to plan for climate change, as Siders's letter of petition had requested. And to this date, no other electric utility in New York has completed a climate assessment resembling what Con Ed achieved. Which means the battle to protect this "vital service" from climate damage must continue—not only in New York, but in other states too.

Not everyone has the time or training to crash a rate case. But there are things you can do. For instance, Siders recommends contacting the utility commission to learn about ongoing cases and to find out what parties are already participating in them. "There is likely be someone who is already representing customers," she said. "You want to find someone to represent you. It's far more effective to have an expert who understands the rate case and all of its ins and outs." And don't forget university law clinics. Over the years, many law schools have developed training clinics specializing in environmental and energy-related issues. There law students and experienced lawyers represent low-income people and community groups for little or no fee. The Con Ed rate case may not have had the success it did without the help of the Columbia Climate Law Center and Pace's Energy and Climate Center. So that resilience advocates don't have to reinvent the wheel, Columbia's center has even published a detailed guide explaining how to use utility commission proceedings to press for sensible climate action.[24] Young activist lawyers, power up!

◆

Enhancing the energy grid with "smart" digital technology and green sources of power would increase the energy network's flexibility and responsiveness. If a tornado knocks out the transformer at a gas plant, a smart grid can swiftly pull more power from the wind farm. When triple-degree temperatures threaten a city's elderly population, computerized relays assign priority to hospitals and cooling stations where electricity is needed most. Smart technology can also allow electricity to move back and forth, supplying a home with power from the gas plant in the evening, and allowing the home's rooftop solar panels to send surplus power to other users during the day. Cities now using smart-grid technology, like Chattanooga, Tennessee, not only report fewer outages but also save customers money over the long haul.[25]

Renewable energy sources don't just reduce carbon dioxide emissions—many appear far less vulnerable to extreme weather. This is because they aren't dependent on external supplies of fuel or water; they rely on simpler mechanical systems that are more compatible with natural processes. That's particularly true of wind and solar. For instance, while broiling in a Texas heat wave in 2011, wind generators helped keep the lights on when several gas and coal plants were forced to shut down. During Hurricane Sandy, grid operators for the affected six-state area received not a single report of damage to the network's many wind turbines and solar facilities. The five-turbine Jersey Atlantic Wind Project, off the coast of Atlantic City, New Jersey, took a direct hit from the storm but was back producing power hours after it had passed. Finally, energy-efficiency programs that reduce electricity demand help lighten the load on the grid, making it more flexible and resilient. Not coincidentally, they also lower carbon emissions, which helps slow global warming.[26]

Smart and green technologies can be combined to create sustainable "microgrids." These are localized grids that are connected to the larger grid, but capable of operating autonomously. In an emergency, when the regional network is disrupted, a microgrid can disconnect from the system and use its own wind turbines or solar panels to maintain a stable (though sometimes reduced) supply of electricity. Microgrids are particularly well suited to facilities that supply critical needs, like hospitals, nursing homes, and evacuation centers.

There are many roads toward a more sustainable grid, but for those looking for inspiration and a practical lesson, I point you to Wendy Gao, who at age fifteen helped "solarize" her public high school in Fairfax County, Virginia. I first learned about Gao while researching the growth of solar energy in school systems across the United States. Enabled by green-energy incentives at the state and federal level, the number of schools powering themselves with installed solar systems has grown by 139 percent since 2014. The technology not only lowers energy costs, but also builds climate resilience by allowing the grid to trade out more vulnerable sources of power. Schools that add battery storage to their solar systems can create their own microgrids, allowing them to run classrooms during rolling blackouts or to provide community shelter during emergencies.

Another thing I learned: amid the scores of efforts to install solar in middle schools and high schools—from Tacoma to Fairfax—it is often the students, themselves, who are leading the charge. The results are stirring. Gao's solar-energy campaign not only transformed her school district, but also, as we will see, her whole career trajectory.

When I caught up with Gao, by Zoom, she was sitting in her dorm room at the University of Virginia. It was the spring of 2021, and she was just finishing her first year of college. Her desk was uncluttered and well-lit. To Gao's right was a cupboard festooned with snapshots. On the wall behind, a watercolor with upwellings of lapis and ultramarine. I asked how she got interested in solar energy.

"It really was just a passion for the planet," she told me. "I had always known about climate change and felt very helpless about it. And I hated that. So, I joined my school's environmental club to see if I could do anything for my community." As part of her involvement, Gao agreed to speak at a small rally in support of the Green New Deal, a set of ambitious climate goals intended to transform the U.S. energy economy. To her surprise, the audience, who included community leaders and a few state lawmakers, was quite enthusiastic. The experience made her feel "the exact opposite of helpless."

"People I didn't know were coming up and telling me how inspiring it was to see young people like me doing this. I realized that my voice was more powerful than I had thought." Gao, who was known among her teachers as a gifted writer, found that her talent transferred easily to public speaking. "I said, you know, this could become a future thing. I could use my words and continue taking this to places like the Virginia General Assembly. And that was the first time I realized that, you know, my voice matters."

As part of another project, Gao and her classmates analyzed their school's electricity use. They discovered that converting to photovoltaic solar panels would not only reduce greenhouse gases and increase reliability, but that it could also save the school district a lot of money. The students were particularly excited about a federal tax credit that at the time allowed a residential or commercial developer to deduct 30 percent of the cost of a solar energy system. But there was a hitch: the Fairfax County School District is not a residential or commercial developer. Besides, school districts don't generally pay federal taxes.

This sent Gao—whose passions were previously environmentalism and writing—down the research rabbit hole of energy finance. What she learned is that it was possible for Oakton High to enter an arrangement with a third-party developer in which the developer would install and operate the campus solar energy system and the school would buy the power from the developer. The developer would take advantage of the tax credit and pass along part of the benefit in the form of a favorable flat rate that would last for the duration of the contract—normally about fifteen to twenty years. This financing mechanism, known as a Power Purchase Agreement, was the key to Oakton's solar puzzle.

With that insight, Gao and her classmates launched a petition, met with local officials, and presented their case at school board meetings. Some adults cautioned Gao to go more slowly, to wait until she was at least a year older to rattle the gates of power. But Congress had designed the value of the tax credit to recede over time. To maximize the benefit, the district had to pounce. "They're saying, turn sixteen," said Gao. "And I'm like, hey, there's an opportunity for you to install solar *now*."

In January 2019, thanks to the efforts of Gao and others, the school district finally agreed to install solar panels at Oakton High and at two other schools as well. By the spring of 2021, the school board had announced plans to use Power Purchase Agreements to make solar energy available to all 87 facilities within the district. (The federal tax credit was extended; and the district was able to take advantage of that.)

Gao continued to spread her wings. She did, indeed, realize her goal of testifying before the Virginia General Assembly, advocating in favor of several green energy bills. She even made an important address at the United Nations Youth Climate Summit. Recently, Gao helped found a youth-led international climate change organization called Earth Uprising, which has a presence on five continents. Gao serves as the lead finance coordinator.

While Gao has become somewhat of an expert in raising and disbursing money, she told me that capital management had not previously been an interest of hers. Her passion was the environment; and if knowing finance was important to protecting the environment, then she was going to learn about it. That's what being forward-looking and flexible is all about.

I asked if she saw any difference in viewpoint or temperament between the advocates of her generation and advocates of previous generations, like me, who see ourselves as fighting for many of the same things.

"Yes, there definitely is. I think that the youth are not as patient or not as tolerant as adults," she said—and I wondered how long she could plausibly exclude herself from that last category. "Adults, I think, are more jaded and a little bit more cynical, and slow to demand change—or to even believe that change is possible. The climate is pretty simple to us. All we want is our lives and our futures."

—————◆—————

At this point, we have seen the risks that extreme events pose to the grid and the damage they can do to communities, particularly those in more vulnerable situations. We've also seen that the path of resilience follows a commitment to hardening structures to make them more secure, smartening the network to make it more flexible and responsive, and greening the generating capacity.

In addition to rate cases and local solar campaigns, we can imagine deeper, more wide-ranging prescriptions. Federal and state regulators surely need to impose stronger mandatory standards to promote hardening, smartening, and greening. The grid is also in desperate need of federal investment. A report by the American Society of Civil Engineers estimates that by 2040, the gap between expected funding and needs will surpass half a trillion dollars.[27]

Congress gave a much-needed jolt to grid modernization when it passed the 2021 infrastructure bill. That legislation invests $65 billion in projects to improve grid reliability and resilience, expand and upgrade transmission lines, improve grid flexibility with computers, and boost cybersecurity. As a result, we can expect fewer and shorter blackouts, cheaper energy, and a less smog. The Inflation Reduction Act, passed in 2022, also improves resilience by focusing on how electricity is produced. Described by President Biden as "the biggest step forward on climate ever—ever," the law raises $369 billion to expand cleaner energy, promote electric cars, and boost the domestic production of batteries, solar panels and wind turbines. The goal is to drive as much renewable development as possible in the most heavily polluting parts of the economy: transportation and electricity generation. A big chunk of that will boost the supply and demand for more resilient forms of power generation like solar panels and wind turbines. Analysts say incentives under the Inflation Reduction Act, spread out over the next decade, are likely to cut carbon pollution by around 40 percent below 2005 levels by 2030, bringing us closer to President Biden's goal of cutting carbon emissions in half within the decade.[28] (Remember, in addition to resilience, carbon reduction is the *other* Job One.)

Investing in resilience is important, but you still need to know how to spend the money. As the Con Ed experience shows, today's network assessments and planning still amount to a piecemeal process. That is not surprising. For reasons discussed in chapter 3, climate resilience planning of all kinds is uneven. Still, there is no way to embark on a path toward comprehensive grid resilience without a clear understanding of what the local and regional risks are and what national and community priorities should take precedence. Researchers at Columbia's Sabin Center report that *no* regional transmission operators in the United States has assessed the ways climate change will amplify extreme events in their jurisdictions. Nor has the Federal Energy Regulatory Commission, known as FERC, made any assessments of climate risks to high-voltage systems. Without such information, anyone tasked with estimating reliability or calculating a reasonable fee for predictable power is shooting in the dark.

As a preliminary step, FERC should team up with regional transmission operators to assess vulnerabilities, identify the main weaknesses to address, and recommend solutions. Like the landmark Con Ed study, these assessments should be based on "downscaled" projections, reflect "worst case" scenarios, and include methods for continually updating information.

States also need to rethink the antiquated cost-of-service rate model, which encourages utilities to pursue expensive capital investments intended to expand geographic access and increase consumption. It's true that such capital investment is increasingly steered into storm fortification, but very little of that includes serious climate considerations. And cost-of-service vastly underrewards investments in smart technology and microgrids. Because these technologies are generally less expensive than old-style infrastructure, the resulting revenue (which is based on the size of investment) can't compete. In general, states that have seen strong growth in greening and smartening have either imposed other policies—green energy standards or resilience requirements—to push the technology forward or modified the cost-of-service mechanism.

Some experts argue that cost-of-service should be scrapped for something called "performance-based metrics." Instead of paying utilities based on the amount of capital they sink into traditional hardware, states could tie revenue to more modern measures, like flexibility, efficiency, and resilience.[29] As we saw in chapter 3, such alignment of interests can be a key tool of resilience planning. Some states, like Illinois and Rhode Island, are now experimenting with innovative pricing models. More should.

Rumbling beneath the challenge of energy resilience is the trauma of energy insecurity. Even in cities with ambitious plans, if the fortifications and smart meters bypass your neighborhood, it does no good. And there's still the monthly bill to pay—the one that a third of U.S. households are struggling with. One day I was staring across Carrollton Street in one of the many historic neighborhoods of New Orleans as an electric streetcar rattled by. These delightful contraptions, with their clanging bells and spearmint paint jobs, have been serving the city for nearly a hundred years—driven by horses, steam, and for most of their history electromagnetic current.[30] I thought of the people I see every day who depend on the city's vital service of reliable and affordable electricity. The owner of the Latino grocery, with freezers full of Rocket Pops, tamales, and butchered pork; the clutch of middle schoolers at the crosswalk, texting updates to their parents on battery-powered phones; and the chatty lady with sequined support stockings who motors her electric scooter across four lanes of traffic every morning to buy a scone and a tall drip at the café where I often write. How do we build energy *security* into *resilience*?

I've been helped in my thinking on this by Shalanda Baker, whose book *Revolutionary Power: An Activist's Guide to the Energy Transition* proposes a "new" energy system centered on "the voices, hopes, and dreams of the poor, people of color, Indigenous people, and those marginalized by the old energy system."[31] At the core is an argument for community ownership and control. Ownership implies the ability to earn revenue, benefit from government subsidies and tax credits, and enjoy other advantages of investment and proprietorship. Control includes the ability to set standards, direct activities, and allocate services. Ownership and control are important in all features of an energy system and, particularly, where resilience is concerned: without good incentives and regulatory control, you can't have much hardening, smartening, or greening.

One day I invited Baker to speak to the students in my Energy Law class; and I was thrilled when she accepted. To my students, Baker explained the importance of ownership and control, noting the misalignment of interests between the parties that generate and distribute electricity and consumers in marginalized communities. Experts have long known that low-income households pay a higher proportion of their monthly income for electricity than more affluent households.[32] There is a similar disparity, Baker told my students, that is based on race and *independent* of income. She noted studies showing "that people of color—even when they're wealthier—tend to pay more for electricity, because of patterns of residential segregation in this country where we have folks living in substandard housing, and electricity is just kind of going out the window."[33]

What is more, technologies like rooftop solar—which promise financial benefits and improved resilience—are all but nonexistent in less affluent areas. The solar tax credits that the federal government and many states offer are "not advantageous for lower- to moderate-income folks, because they don't have the tax liability." As with affordability, the solar divide is about more than income. "Even when you control for homeownership rates, and even when you control for income," Baker explained, "White communities tend to have more solar." Again, the phenomenon relates to racialized housing patterns. The infrastructure and marketing of more sustainable resilient power systems are generally directed toward communities that are mostly white. Baker takes a long pause while the point sinks in. "That's a little bit mind blowing, right? It essentially means this solar transition is happening along racial lines."

Part of the solution might involve more accessible power purchase agreements, like the one used by Wendy Gao's high school to capture part of the benefit of solar tax credits. That would help families who can't afford the upfront installation costs or who don't pay enough in income taxes to take the full deduction. The deeper, more revolutionary move is community ownership. Baker argued that more municipalities should take ownership of their electricity networks, operating them as nonprofit enterprises and managed through a politically accountable board. Customers of publicly owned utilities tend to pay lower rates than customers of privately owned utilities, and experience shorter power outages. Also, the revenue from a publicly owned utility is more likely to be invested in the local community.[34] My faith in public ownership is perhaps not as vigorous as Baker's; but in a country with over 70 percent private ownership, there is surely room for more experimentation. And we both agree that community ownership of microgrids is a promising way for low-income communities to strengthen their local economies and build resilience. "I mean, there are a lot of different things that can come out of being able to say I own this, and I'm going to get the benefits and I get to control how those benefits are distributed," she said.

Baker also believes community advocates should not be afraid to dive into regulatory proceedings the way A. R. Siders did. Baker recommended that state utility commissions be urged to adopt ratemaking formulas that favor green development and pollution reduction in neighborhoods that were underserved and overcharged. I said those formulas should encourage climate resilience, too. She agreed, adding, "this is a new frontier in energy policy research."

Baker spoke to my class at Loyola on November 30, 2020. Coincidentally, the date marked the last day of the Atlantic hurricane season, the one that set the record for the most named storms in a season—thirty of them. That day, yellow-vested workers in the southwestern Louisiana were still laboring to rebuild a swath of the electricity grid that Hurricane Laura, the strongest hurricane to hit the state in 164 years, had demolished in late August. The devastation had left 360,000 customers stranded without power and water for days. A spokesperson for Entergy said at the time that the damage to high-voltage lines and other installations was some of the most severe the company had ever experienced. Some of the places, reported a journalist from CNN, "would need to be rebuilt from the ground up."[35] Even as the heat index hit triple digits and clouds of mosquitoes choked the sky, thousands of residents reemerged, some hauling newly duct-taped refrigerators to the curbs. What inscription, I wondered, would I have painted on one of those mildewed monuments?

Forget funky. Smells like climate breakdown to me.

chapter 8

Flash! Crack! Boom!

Thirty miles north of Seattle, between the Olympic Peninsula and the city's metropolitan corridor, Whidbey Island marks the northern boundary of Washington's Puget Sound. The island's rugged terrain holds hills, forests, and cliffside beaches. It's a wonderful place for sea kayaking, something I do there every summer. My favorite spot is a small indent on the southwest shore called Useless Bay. The name is attributed to a nineteenth-century U.S. survey expedition, which found the shallow cove poor shelter for storm-threatened ships.[1] Of course, the Coast Salish peoples had, for centuries, already found the bay plenty useful as a site for fishing, crabbing, trading, and even recreational paddling—but that's another story.

In 2020, on Labor Day morning, I remember slicing through the water in my yellow kayak. It was quiet and still, but things were not right. The eddies from my paddle—which normally sparkled silver and blue—were instead a ruddy lather. The surrounding bluffs were flat black, not umber, not green. The sky overhead was bright orange, and yet somehow without sun. It was a tableau worthy of comparison to Edvard Munch's painting *The Scream*, I remember thinking, that surreal cliché of existential panic.

Except I wasn't panicked, or, at that point, even surprised. For this was the weekend of the so-called Labor Day Fires, the enormous wildland blazes that were ripping through inland forests of Washington, Oregon, and northern California. There was no risk of forest fire on Whidbey Island or in the Seattle metropolitan area, but the sheets of smoke and gas wafting in from miles away were penetrating. These lenticular layers drifted hundreds of feet above, filtering out the blue morning light and allowing only powdered ash to rain down. For that reason, I was thankful to have a KN95 face mask, which I normally used for weekly grocery shopping during the pandemic. Still, my eyes stung a bit when a breeze kicked up. A flock of terns cut through the haze, and I wondered what they made of all this.

I pulled a Ziploc bag with blackberries from the pocket of my life jacket. Earlier that morning I had picked a hatful of those berries while walking along the road. I raised my mask and popped a couple in my mouth. There was a sweet burst of flavor, and then an aftertaste of dust.

The orange haze hung over Useless Bay for more than a week.[2] I checked the satellite map on my iPhone several times a day, but the image was always the same: a 1,500-mile-long trail of smoke, resembling a Nike swoosh, its head swirling over Washington, Oregon, and northern California and its tail stretching toward Hawaiʻi. The effect, I learned later, was caused by a sharp drop in atmospheric pressure along the coast. The prevailing easterly winds had raced to fill the void, topping fifty miles an hour, and were then shoveling mountains of smoke across the burning Cascade Range and into the Puget Sound. During that time, it is said the Seattle metro area had the worst air-quality readings anywhere in the world.[3]

Meanwhile, east of the Cascades, these wildfires were exploding at will across more than a million acres in Washington and Oregon. The perimeters of two of the largest infernos, the Riverside and Beachie Creek fires, stretched more than eight hundred miles— about the length of a one-way trip from Portland, Oregon, to Salt Lake City, Utah. Oregon's Department of Forestry reported that 7,500 firefighters from thirty-nine states had been deployed to contain the firestorm. More than a week later, most of the Labor Day fires would be brought to heel by coastal rains, but not before claiming eleven lives.[4] In addition, more than half a million residents in Oregon and Washington would be forced to evacuate, more than four thousand homes would be razed, and 1,600 square miles of towering cedar, fir, and lodgepole pine would be laid to waste.[5]

Reciting these numbers, I admit I have a hard time comprehending the scope and effect of fire on our landscapes. Wildfires can be fiercely destructive, but they also provide benefits. They are a dramatic symptom of the unspooling climate, but also an important part of the natural order. And since the first time I saw *Bambi* at the movie theater—my fingernails buried in my mother's arm—the *Flash! Crack!* and *Boom!* of forest fire has rattled my nerves like a cymbal crash.

Tadd Perkins, a friend who lives in Seattle, spent a career fighting fires of many kinds, in forests, suburban neighborhoods, and industrial facilities. Fire-fighting runs in the family and both his brother and his son, Sam, share the profession. When I first met Sam, he was five years old, about the same age I was when my mom took me to that Disney film. Now in his mid-twenties, Sam is a firefighting specialist known as a "smokejumper." He parachutes into wildland fires to keep them from spreading.

Sam trained as a volunteer firefighter during college, just as his dad did. Upon earning his degree, Sam joined the U.S. Forest Service to develop his expertise. It was a way, as Sam saw it, to merge his love of the outdoors with a deep commitment to public service. Sam has helped battle many of the fires you've probably read about or seen in the news. In fact, as it turns out, Sam had been battling the megafires in Oregon on the very day I was paddling through the choking fog on Useless Bay.

I hadn't seen Sam in years. But when I learned of his role in the Labor Day fires, I knew I had to talk to him. Weeks after the fires subsided, I scheduled a meeting with him over Zoom. Sam had the same broad smile I remember, now framed by a neatly cropped beard. He looked fit, of course, and conveyed the placid demeanor of a long-distance runner at rest. I asked him what it was like to be a smokejumper fighting one of Oregon's most monstrous fires.

"For me, what it might look like is parachuting into the fire and working a long shift or two, followed by a thirty-six-hour shift with no sleep." He tells me it's dangerous and grueling. Jumpers typically parachute in groups of ten or so near what's called the "heel" of the fire, the slowest moving part, near the fire's origin.[6] Once they've landed, the aircraft circles back to drop a load of cargo consisting of all the things they'll need potentially for the next seventy-two hours on their own: food, water, sleeping bags, chainsaws, and specially designed axes that can also be used as hoes. Their first job, which is called the "initial attack," is basically to rip, hack and dig, at near-lunatic speed, until the blaze is separated from its fuel source and gradually dies out. When the job is done, the jumpers each pack up about a hundred pounds of gear and hike several miles to the nearest road for pick up.

Sam's father, Tadd, as I mentioned, also fought wildfires, though not by parachute. I asked Sam how these kinds of fires had changed since the 1970s and '80s, when his dad was on that beat.

"A big fire for him was ten to twenty thousand acres. In the past few years, I've been on numerous fires that have been larger than 500,000 acres. I've been on a million-acre fire," he says matter-of-factly.

While jumpers specialize in the "first attack," any change in the situation—a burst of wind, a rancher's call for help—can reshuffle plans and reassign roles. Thus, Sam has also worked the community end of the fire, helping people evacuate and find cover. He told me about a recent blaze in Idaho, where some of the strongest winds he had ever seen swept the fire toward a rural town. "Dealing with the evacuations, dealing with the livestock, that was challenging. The first shift was over forty hours, and the second shift was just shy of that." There's too much going on to focus on the danger, he told me. But no one's oblivious, either. "We've lost colleagues and peers to fires. Every year approximately eighteen firefighters die in the United States."

The work, of course, takes a huge toll on body and mind. He told me about the six-month grind in the field, the stressed joints and "workplace injuries."

"I know plenty of folks in their thirties," he told me, "who have new knees or new hips or have dealt with arthritis or plantar fasciitis or spinal issues." He told me about the prevalence of post-traumatic stress and the quiet surge of suicides among wildland fire fighters over the last decade.[7] The whole scene, as Sam describes it, is surreal and disorienting. I can't imagine. The only window I have on his world is perhaps the days I spent under an ochre sky in Useless Bay.

"That is our life for the summer," he says. "We live in those thick areas of smoke, deal with the poor air quality, the headaches, the dehydration that may occur and perhaps downstream impacts of smoke exposure." Then after some reflection: "But you know, that's our job."

———————————————

Wildfires are the latest cataclysm to be amplified by climate change, burning rapidly and haphazardly across the American West and in other parts of the country too. Too many of us are in the line of fire. We've lost homes and businesses; vineyards and redwoods; livestock and fish. We've lost sisters, brothers, friends, and parents. Winds can carry wildfire smoke hundreds, even thousands of miles, swathing cities in ribbons of soot, tar, and acid gas. The amount of land consumed by wildland fires in the United States has grown steadily for more than sixty years.[8] In fourteen of seventeen years between 2000 and 2016, wildfires devoured 3.7 million acres of land, an area larger than the state of Connecticut.[9]

Size is but one characteristic in the typology of fire. There are "crown fires" that roar through a forest's highest canopies, asphyxiating everything below, "ground fires" that smolder through layers of leaves and peat; "fire tornadoes" that rocket thousands of feet into the air, propelled by 100-mphwinds; and "megafires" characterized by their blinding intensity as well as breadth of damage. Experts recently coined the term "gigafire" to describe a blaze that claims more than a million acres of land. In 2020, California's August Complex fire met that grim definition.[10]

The western states, by the way, are not the only tinderbox in America. The southeastern United States has also seen increased fire outbreaks and a longer fire season.[11] In the next forty years, forestry experts expect the annual area of lightning-ignited wildfire in this region to jump by 30 percent. The southern Appalachian region, whose forests have been degraded by insect infestations, disease, and logging, is particularly vulnerable. In 2016, during a prolonged dry period, fifty major wildfires broke out in that area, burning more than 100,000 acres in eight states and killing fifteen people. It was the worst firestorm Appalachia had seen in more than a hundred years.[12]

The economic burden of so much fire can be severe. From 2002 to 2011, the insured losses related to wildfire totaled $7.9 billion, more than three times those of the previous decade.[13] Since 1960, it's estimated that property loss from wildfire has increased more than 220 times.[14]

Then there's the cost of defending that property by fighting the fires. Sam Perkins gave me some idea of the human cost involved in that endeavor. In terms of money, now the federal government often spends about three billion dollars each year to fight forest fires.[15] Just twenty-five years ago, extinguishing fires—what experts call "wildfire suppression"—made up only about 16 percent of the Forest Service's annual budget. Today, the agency spends more than half of its budget on putting out fires.[16] When suppression costs go over budget—as they often do—the Forest Service is allowed to dip into its other accounts; but those funds aren't replenished. Which is to say that every extra dollar spent for fire suppression is a dollar that is *not* spent on something else we need, like fixing wetlands, fighting bark disease, or, more important, clearing breaks to prevent more forest fires. It's harder to get a picture of what the states together spend on wildfire programs, but a survey from 2014 put the collective figure at almost $2 billion.[17]

Then there's the work of community recovery. Three months after the Labor Day fires, I checked the local news sites to see how hard-hit communities were getting along. Hundreds of Oregonians, I learned, were still waiting on relief they said had been promised. The Federal Emergency Management Agency (FEMA) was then struggling to place another 267 families in mobile homes, though the families' utility bills would still have to be paid out of pocket. In November—just before the dead of winter—it was reported that Oregon lawmakers had agreed to set aside $65 million in state funds to convert motels into shelters for wildfire survivors and people experiencing chronic homelessness.[18]

On top of that, nature itself takes a blow. Countless animals die in wildfires, although many species have learned to cope. Birds fly. Bears run. Frogs slide into the peepholes of ancient logs. Scientists have no good estimates on the number of animals lost to wildfire in the United States, but there seems to be consensus that even species that have evolved to thrive in tinderbox-habitats face higher risks today.[19] During the peak of the epic wildfires that hit Australia in 2020 (called "bushfires" on that continent), scientists estimated that more than one billion animals had perished in New South Wales and Victoria alone.[20]

In addition, wildfire destroys species habitat and can disrupt important ecosystem functions. In many cases, that damage is offset by accompanying ecological benefits. What is not offset are the hundreds of millions of tons of carbon dioxide that forest combustion adds annually to the atmosphere. Every year, U.S. forests breathe in several hundred megatons of carbon dioxide, which are then stored in leaves, branches, and roots. When trees and other forest vegetation burn, that gas zooms back into the sky, where it adds to global warming.

The effect is profound. The California fires in 2020, for example, are thought to have pumped ninety million metric tons of carbon dioxide into the air. That's a third more carbon dioxide than the annual discharge of California's entire electricity grid.[21] Think of all the bold efforts Californians have made to cut carbon over the years—caps on greenhouse gases, leaps in building efficiency standards, tens of billions of dollars in private and public investment poured into fast trains, electric cars, and solar farms.[22] All that progress can disappear in a puff of smoke. All right, gargantuan billows of smoke, if truth be told, whose even short-term effects entail more than you might know.

◆

Cinthia Zermeño Moore thinks a lot about that smoke. Moore is a part-time realtor in Las Vegas and a field consultant for the Moms Clean Air Force, a nationwide organization that advocates for environmental protection and children's health. Moore and her family relocated from North Hollywood during the Vegas construction boom of the 1990s.

Like many American cities, Las Vegas has gotten hotter in recent years, and what little rainfall it once received has diminished. Moore and I both grew up swimming in and hiking around Lake Mead, a breathtaking water body on the Colorado River and just outside the city. Moore worries that today's children won't have the chance to enjoy it the way we did.

In an interview, Moore told me about a time she took her son, Liam, then three years old, to Lake Mead as a way to have some fun during the pandemic. "I was surprised to see the water levels were extremely low," she said. "I remember back then the water levels were low too, but now it's even worse. It really has me worried because at this point, I don't think I'm going to be able to share my experiences that I had at Lake Mead with my son."

As a Las Vegas native, I can relate. Back in the 1980s, when I was in high school, my friends and I would sometimes cut school and sneak out to Lake Mead. We liked to climb a towering shoreline crag we had discovered, which we called "the Cliffs." There we would bake in the sun before taking running jumps into the glassy water below. (I later learned that in the 1950s my mother and *her* friends did exactly the same thing. Is anything truly discovered?)

A few years ago, during a visit with my mom, I had the chance to visit that old shoreline. Today if you jumped off the Cliffs, you'd fall twenty feet onto a crop of boulders. The water's edge is half a football field away. I read later that Lake Mead reservoir currently holds less than 30 percent of its original capacity. These days, the city can now go up to 150 days without any measurable drops of rain.

But I had not met with Moore to talk about Lake Mead. I wanted to talk to her about something hundreds of miles away—the California wildfires. For though she was not feeling the heat from that storm, she and her family were certainly bearing the exhaust. Winds were carrying the smoke across state lines, pumping tons of soot into the already polluted skies of the Las Vegas Valley.

"My son is so young, I feel like he's missing out a lot on his childhood," Moore told me, "I mean, he's only three, but I feel like he's missing out on a lot of things." She's mainly concerned about the city's increased summer heat and the air quality. But when we spoke, she was preoccupied by California's burning forests. "I was driving the other day and I was like, Oh, my gosh, it looks so different. And then I realized that I couldn't see the mountains because of the smoke," she told me. "I thought about taking him to Gilcrease Orchards," she said, referring to a spring-fed "u-pick" farm that is popular with Vegas residents. But when an air hazard warning pinged her cell phone, Moore reconsidered. "Outside, it looked cloudy, but it was *really* smoke, and I was like, I'm not taking my son out there. He has extreme allergy issues."

Moore's fears are backed by the science. In addition to carbon dioxide, wildfires produce two other things that, despite their technical names, you will definitely recognize: ozone and particulate matter. Ozone, or O_3, is a pale gas that smells a bit like chlorine and stings your eyes. Miles up in the stratosphere, ozone does us a favor by forming a protective shield—the "ozone layer"—that screens out much of the sun's harmful radiation. But when ozone occurs at ground level (a result of smokestack and tailpipe exhaust) it causes big problems by turning the sky brown and attacking our lungs. That's when we call it "smog." Lung disease, heart disease, and nervous system disorders are all linked to smog exposure. And, as with so many pollution hazards, the risk is much higher for the young, the elderly, and those with severe allergies or other respiratory problems.[23]

Particulate matter refers to any tiny liquid or solid particle that hangs in the air. It could be soot, pollen, sea spray—almost anything. Many times smaller than a grain of sand, these particles, when inhaled deeply at high concentrations cause serious harm, being associated with lung cancer, heart disease, premature births, respiratory trouble, and many other problems.[24] Of particular concern are a category of particles that measure no more than 2.5 micrometers in diameter. Known as "fine particles," or $PM_{2.5}$, these tiny specks can penetrate deep into the lungs and even the bloodstream. According to the U.S. Centers for Disease Control and Prevention, $PM_{2.5}$ poses the largest environmental risk in the United States and is responsible for more than 100,000 deaths per year. That's more than from homicides and car accidents combined.[25]

The share of ozone and $PM_{2.5}$ attributed to fire is substantial. According to one government report, 40 percent of all $PM_{2.5}$ emitted in the United States in 2011 arose from prescribed burns or wildland fire. The report predicted that as industrial sources of fine particles are further regulated, wildfires will become the main summertime source of this most dangerous pollutant.

I asked Moore if her friends and neighbors shared her concern about the city's whisky-colored skies. "I live in the area where it's highly Latino. I talk to my neighbors a lot," she told me. "Because of the smoke and the bad air quality, they've had sore throats, they've had watery eyes, they're sneezing, they have headaches. At first, they were afraid because they thought it was Covid. I started talking to them about the air quality and if they made that connection."

It was a relief, sort of, but now their conversations have turned to asthma, which can be seriously aggravated by both ozone and fine particles. "The Latino community is one of the hardest-hit communities here when it comes to high asthma rates," Moore said, "and we have one of the highest rates of kids missing school because of asthma attacks." When the smoke came, Moore noticed that Liam's allergies began to flare and that he had some trouble breathing. When Liam broke out in a nasty rash, she invested in a portable air purifier. "I don't want to just give him allergy medicine all the time," she said.

Working backward from Liam's air purifier to smog and soot to Sam Perkin's brutish tours in hellfire, it's possible to trace the root of our new wildfire crisis, which, I learned after considerable investigation, are actually three intertwined roots: climate disruption, bad forest management, and a residential fantasy known to planners and firefighters as the "Woo-Eee."

There's little doubt that the climate crisis is intensifying fire behavior and expanding its range. As the climate unravels, warmer springs, longer summer dry seasons, and drier soils have all lengthened the wildfire season.[26] In the West, long dry spells are often followed by intervals of intense downpour, a kind of "weather whiplash."[27] The rain feeds the growth of thick underbrush, which dries out in the next drought, which becomes a flammable raceway for the next megafire.

On account of climate change, according to a recent government study, the area of burned forests in the United States from 1984 to 2015 actually doubled.

The study predicts climate change will continue to multiply wildfire frequency and burned area well into twenty-first century.[28]

Take note that I am *not* saying the Labor Day fires or California's August Complex fire were caused by climate breakdown. There were lots of overlapping causes, from climate cycles to forest-management techniques. And don't forget that humans are responsible for nearly 85 percent of wildfire ignitions.[29] Downed power lines, a topic we considered in the previous chapter, are thought to have sparked at least a few of the Labor Day blazes. Our confidence in attributing a given event to climate breakdown will vary by the kind of event. Scientists are generally much more confident about associating climate change with heat waves or cold snaps, for instance, than with a wildfire.[30] The more helpful question is not "Did we *do* that?" but, "Are we *encouraging* that?" And if so, "By how much?"

On this score, scientists in the field of extreme-event attribution have a lot to say on the subject of wildfire. In one technique, scientists take observations from a real-world fire event and, using computer models, compare those observations to a similar " counterfactual" fire in which climate change does not exist. The variance in strength and damage reflects what we might call "the climate change difference."

Using methods like these, scientists with the World Weather Attribution research consortium examined the massive bushfires that took place in Australia in 2020. They found that climate disruption increased the likelihood of the necessary conditions for such a blaze by at least 30 percent.[31] Another research team looking at British Columbia's record-breaking fire season of 2017 found that climate disruption made the conditions that fueled those fires two to four times more likely and expanded the burned area at least sevenfold.[32]

One factor that has turned out to be critically important in analyzing the effects of climate change on wildfires is something called the vapor pressure deficit.[33] This is the difference between the amount of moisture in the air compared to the amount of moisture the air could hold if fully saturated. As air warms, its ability to absorb water increases. A high vapor pressure deficit indicates air that is very dry and eager to sap moisture from surrounding vegetation like leaves, shrubs, and grasses. When the vegetation dries out, it's fuel for fire.

A 2019 study in the journal *Earth's Future* found that among the many drivers of California wildfires, "warming-driven fuel drying is the clearest link between anthropogenic climate change and increased California wildfire activity to date."[34]

Studies show that fuel-drying vapor deficits have been climbing in California and other western states for generations. The summer I paddled under Washington's orange skies, the vapor pressure in Oregon and California had hit levels that had not been seen in forty years.[35]

It's important to keep in mind that wildfires are a normal phenomenon across many parts of the United States and the world. Carbonized cedar, fir, and aspen all have an afterlife. Their remains return nutrients to the soil and provide feeding and nesting grounds for a variety of species, including mule deer and woodpeckers. Wildfire beats back insect infestations, promotes wildflower diversity, and awakens new generations of lodgepole pine by scattering its seeds. It also clears out debris (leaves, limbs) that, left unchecked, would feed blazes too big to be considered healthful. Indigenous peoples understood this, and for centuries integrated prescribed burns into their land-management practices.

But somewhere along the way, our tolerance for natural fires gave way. Fire historians trace that change to a wind-driven conflagration in 1910 called the "Big Blow Up," which raged through Idaho, Montana, and Washington. An inferno made up of hundreds of separate fires, the blaze charred more than three million acres and killed eighty-five people, including seventy-eight firefighters. Only five years in its existence, the Forest Service vowed nothing like that would ever happen again. The agency invested heavily in firefighting capability and adopted an aggressive policy of fire suppression that continues today.[36] Because of that policy, forests are now chock-a-block with countless tons of flammable material—dried-out logs, withered brush, powdered leaves. Researchers estimate that California's forests are filled with more than 129 million standing dead trees, all waiting like roman candles for the next summertime spark.[37]

Accounts like these have opened new conversations in the West about "forest management," by which is meant some combination of logging, prescribed burns, and underbrush removal. President Donald Trump championed logging as a promising way to prevent wildfire, as did President George W. Bush.[38] Today, the state of Oregon is also looking at ways to reduce wildfire through more logging and other forms of forest management over the next twenty years.[39]

Forestry experts have their doubts. Oregon's westside forests, for instance, are so vast that treatments at such scale may be ecologically impractical and unaffordable. And they could make things worse. The forests destroyed in Oregon's Labor Day fires were, after all, anything but untouched. The blazes fed on some of the most heavily logged timberlands in the state About 43 percent of the land within the perimeter of the Labor Day fires was privately owned, managed by companies like Weyerhauser, Freres Lumber, and Seneca Jones Timber.[40] The bulk of the remainder was federally owned and managed for harvest by the Forest Service and the Bureau of Land Management.[41] This was not original growth, but replanted forests in their second and third generations.

That turns out to be part of the problem. The United States is one of the most forested countries in the world; but its hair is overstyled.[42] Just over half of our forests are privately owned, and two-thirds of all forested area is classified as commercial timberland. Over generations, the timber industry not only shaved away most of the nation's forest cover, its dramatically reshaped what we have left.

Before European settlement, North American forests were varied and diverse, with alternating patches of trees and scrub that served as fire breaks. When natural wildfires erupted, a necessary part of the regenerative cycle, they usually burned out quickly.

Industrialized timber practices changed that through a series of techniques, most notably clearcutting. When acres of forest are lopped away through clearcutting, they are replanted with just a few species of trees, depriving the land of variety and natural breaks. These new forests, as uniform as a book of matches, are especially vulnerable to fire.

Let's get back to the Labor Day fires. Recent evidence shows logging was *not* the firefighter's friend. In one of those blazes—the Holiday Farm fire—infrared data gathered by Forest Service aircraft showed fires blasting through logging corridors as if from a flamethrower. Satellite data from the same event confirm that the fire burned hottest and fastest in privately managed timber plantations. "The fire areas had been logged left, right, and center and it did not stop the fires," said Erik Fernandez, a wilderness program manager, of the Labor Day fires. "If anything, it may have fueled the fires even more."[43]

Researchers still say there is a role for thinning trees and clearing brush—what the Forest Service calls "treatments." But most experts acknowledge this as a limited response, perhaps best used near populated areas. According to one study published by the National Academy of Sciences, only about 1 percent of land treated by the Forest Service experiences fire in an average year.[44] Plus the protection is only temporary. "The effectiveness of forest treatments lasts about 10–20 years," the study concluded, "suggesting that most treatments have little influence on wildfire."[45]

A solution that might make sense is allowing moderate fires to clear out crowded woodlands, either by letting existing fires eat through the understory or by launching prescribed burns. Native tribes in the Pacific Northwest have used fire as a forest management tool for thousands of years. A touch of flame promoted the growth of huckleberry and the proliferation of the lily-like camas plant, a food staple of coastal tribes. Low-intensity blazes cleared the understory of thorny stalks and poison oak for easier passage. The flavor of the charred debris attracted deer and elk for easy hunting.[46] These controlled burns, introduced on a rotating basis at least every ten years, also helped prevent catastrophic fires.

Cheryl Kennedy, chairwoman of the Confederated Tribes of Grand Ronde, says that such burns are also an important part of Indigenous culture. "Done in a controlled way, as a management tool, as opposed to out of control, fire improves the resources the creator has given us," she told a reporter for the Portland *Oregonian*. "This knowledge was instilled in us and we look forward to the day when these practices are as widely used as they should be."[47]

Compared to logging and mechanical clearing, prescribed burns are a financial bargain. A ten-thousand-acre prescribed burn in Oregon's Fremont Winema National Forest, for instance, cost only about $20,000.[48] Even so, public agencies still lack the resources to introduce controlled burns. The Forest Service, remember, has to dip into its other budgets just to fight ongoing fires. Another concern is the serious air pollution those fires produce. In Oregon, state and federal pollution standards essentially bar controlled burns in any place that could send smoke into a populated area.[49] Thinking back to Cinthia Zermeño Moore and her son Liam, it's not hard to see why.

In addition to the climate disruption and bad forest management, there is a third intertwined cause of the today's wildfire crisis, which Sam Perkins spelled out for me in an almost offhand way. "The truth is," he said, "we live in areas that are *designed* to burn." He was referring to the fact that, since the 1940s, so much of the country's residential development has stretched toward naturally fire-prone landscapes, like meadows, grasslands, and forests. This zone, where natural areas and development meet, is known to planners as the WUI, for Wildland-Urban Interface, pronounced "Woo-Eee." The WUI often presents as an idyllic backdrop for easy suburban living. But underneath, it's a stack of kindling.

When geographer Stephen Strader examined the history of wildland fires from 1940 to 2010, he found that a significant share of increased property damage from wildfire was attributable to what he called a larger "human-development footprint." Fueled by a hot postwar economy and a new interstate highway system, footprints like these expanded in the Southeast and in the West like famished amoebas. According to Strader, from 1940 to 2010, the number of homes built in natural fire-hazard zones increased by nearly 135 times.[50] Woo-Eee.

Perhaps no transformation was more dramatic than residential expansion on the West Coast. Today, in Washington, Oregon, and California, more than 4.5 million homes are located in areas with high wildfire risk. Together, the structures are valued at about $3.3 trillion.[51] People I talk to sometimes assume the WUI is populated with the new elite, with cannabis moguls and media influencers enjoying views of Mount Hood or Lake Tahoe from cantilevered decks. There's some of that, for sure. But for the most part, the WUI is not nearly so fancy. It's like a lot of places you've seen, with tract homes, Panera cafés, and multiscreen movie theaters. In some regions, it appears, people choose the WUI because affordable housing in lower-risk areas is not available.

This is the conclusion of a recent report conducted by Redfin, the popular online, technology-powered real-estate brokerage. The report, which examined residential properties in Washington, Oregon, and California, finds that double-digit price growth in some West Coast cities is driving homebuyers from pricey markets like San Francisco to more affordable, fire-prone areas, like Santa Rosa or Sacramento. The study notes that since 2010 home values have risen almost 10 percent more in zip codes where wildfire risk is low than in zip codes where wildfire risk is high.[52]

Redfin's chief economist, Daryl Fairweather, is concerned. "The lower cost of housing in wildfire-prone areas compared to low-risk areas is likely just the beginning of the consequences of climate change for the housing market," she said, on the day of the report's release. "Right now, wildfires are still a rare occurrence for homeowners, but if fires and other climate disasters continue to happen more and more frequently, some housing markets will go from less desirable to untenable, yet they will remain the only option for many families."[53]

When will the signal be heard? People are still migrating to California's Central Valley and Oregon wine country because, to many, the risk still seems remote. But a new map-based website created by the Forest Service might change that. Called "Wildfire Risk to Communities," the website provides an interactive set of maps and charts "to help communities understand, explore, and reduce wildfire risk."[54] Funded by Congress in 2018, it's the first tool to provide mapped data about community wildfire risk on a national scale with uniform measures.[55]

I'm a big believer in maps and other graphic means to identify and describe risk. The week the "Wildfire Risk" website went live, I spent hours tracing the contours of the nation's wildfire bull's eyes. I even showed my wife. One evening, we gazed together at the orange-red bands undulating over the coastal and interior forests of the Pacific Northwest.

"They spent taxpayer money just to show where the *forests* are?" she asked.

You might be thinking this too. And if you are, I get it. But let me tell you about the *other* information that is folded into these displays. In addition to facts about vegetation canopy and land cover, the maps incorporate temperature and wind patterns over recent decades, and instances of wildfire occurring up to 2015. Significantly, these data points all reflect recent trends in climate change. The maps also include information about the *human* environment, like housing density, housing type, and demographic information taken from the U.S. Census. The point is not just to plot out where a wildfire might *occur*, but to understand what human harm that fire might *cause*. Forest Service staff hope to update this information at regular intervals, although, as the website laments, Congress has yet to appropriate money for that.[56]

Let's take it for a test drive. Imagine you want to know something about the risk faced by the community of Paradise, in Butte County, California. This postcard town in the Sierra Nevada was incinerated in California's 2018 Camp Fire, an event which remains the deadliest and most destructive wildfire in the state's history. You would type in the words, "Paradise, Butte County." A color-coded map would appear with a column of charts and statistics, reflecting, in this case, conditions as they existed just months *before* the Camp Fire occurred.

You would learn that, at the time, Paradise faced a fire exposure risk that was "65 percent greater" than other communities in California. A large majority of the risk came from "direct sources," that is, "adjacent flammable vegetation." You would also see that as of that date a quarter of Paradise's 26,000 residents were over sixty-five years old and that nearly as many had a disability of some kind. Seventeen percent of the population (1,850 people) lived in mobile homes, which, you would learn by clicking a tab, are more likely than houses to collapse during wildfires and therefore pose greater risk to their occupants. Finally, you would see that at the time before the Camp Fire struck, 7 percent of all households (778) had no vehicle with which to escape.[57]

The data are not precise enough to measure risk at the scale of an individual home or even a neighborhood. But the Forest Service believes the tool can help state and local governments to identify vulnerable communities and deploy their financial and regulatory powers to dampen the risk of searing catastrophe.

Walking along a road one winter day in 1892, the Norwegian painter Edvard Munch was seized by a "wave of sadness," accompanied by a reddening of the atmosphere and an eruption of "flaming clouds like blood and swords."[58] Short of breath, he stopped to lean against a fence. His companions kept walking. "I stood there quaking with angst," Munch wrote in his diary, "and I felt as though a vast, endless scream [had] passed through nature. [59]

Meteorologists have long wondered what might have caused those broiling skies that so impressed Munch. Some opined that it was smoke and gas wafting from an explosive volcano, Mount Awu in North Sulawesi, say, or legendary Krakatau. A study in the *Journal of the American Meteorological Society* speculates that it may have been the ghostlike ribbons of "polar stratospheric clouds" that made Munch's stomach drop.[60]

Years ago, during a visit to Oslo, I ducked out of an academic conference and caught a cab to Norway's National Gallery, where the best-known rendering of *The Scream* is on permanent display. (There are at least four versions.) I wanted to stare into the face of that suffering figure—drained of sex, race, and even language—as he or she covered his or her ears in a vain attempt to smother nature's primal shriek.

And, as I stared into that abyss, I recall being a little disappointed.

The painting was smaller than I had imagined and, to be honest, seemed a little haphazard in execution. On close inspection, I could see that beneath all the pain and terror that walloped me on first sight were just grainy streaks of tempera on unprimed cardboard.

I'm not criticizing—add visual art to the list of things on which I have no pre-scribed training. But the experience calls up what has become one of this book's central themes: Within the wallop of pain and terror that so many of us feel in this carbon-drenched epoch, there can be relief when one chooses to stop, inspect, and reorganize one's perception. It's true that climate change is a lot *more* than cardboard and paint. But it need not paralyze us.

When Sam Perkins jumps from a plane into whirl of gas and soot, his focus, he told me, keeps fear at bay. When Cinthia Zermeño Moore troubleshoots her son Liam's allergies or defends her polluted neighborhoods in local hearings, she is choosing action over gridlock, agility over angst. What would it mean for society to do that?

Resilience work means reaching *beyond* carbon dioxide and other greenhouse gases. We know climate disruption is amplifying the wildfire crisis and that cutting carbon emissions is paramount. But that can't happen fast enough, so we must also minimize the alternative contributors to wildfire hazard, in particular the contributors that we have more control over and that we can address more quickly. That means improving the ways our forests are managed and the ways our communities are built.

Most fire experts agree that the nation's forest-management strategy, particularly in the western and southeastern parts of the country, needs an overhaul. Suppression tactics should be limited mainly to protecting human settlement or respiratory health. Otherwise we should let public forests burn in order to clear out the clutter and reduce the risk of larger, more damaging fires. This will require a cultural shift among some forest managers and a shift in public expectations. Many of us who hike or hunt or fish in public wildlands don't like traipsing through broken, charred landscapes. We bought the kids a new tent and drove a hundred miles for *this*? Yes, in fact, you did. And there's plenty in that ash to marvel at and learn from. Culture can change quickly if we want it to.

Prescribed burns have a place too, even though their benefits are limited in scale. Such fires make particular sense near populous communities, though that again raises the issue of respiratory health. I would trust that a decision-making process involving fire experts, public health officials, and informed community groups should be able to balance the risks in such situations. More prescribed burns will require more resources, too, a problem that could be solved by providing a dedicated budget commensurate with the challenge of fire prevention.

The harder, and arguably more urgent, issue involves residential development. The WUI bull's-eye is expanding quicker than the flaming arrows are flying. But we don't address the problem because our incentives are all wrong. State and local governments lack proper incentives to regulate building and land use in the WUI, and homeowners lack information and proper incentives to reduce fire risk on their property.

At the state and local level, there's a mismatch in cost-sharing. When fires cross political boundaries, say between state and federal lands, the federal government picks up most of the tab, regardless of where the blaze started. Local, county, and state governments receive all the benefits of new development in the WUI (from property taxes, building permit revenue, and the rest), without paying their share of back-end fire-suppression costs.

The result, according to Timothy Ingalsbee, a fire researcher at the University of Oregon, is that "taxpayers across the country are essentially 'subsidizing' private development in an expanding WUI by providing free/low-cost fire protection to private property owners."[61] Meanwhile, local governments are *encouraged* to expand development in the WUI because it looks like an easy way to provide housing and increase revenue from property tax.

Most western states have some laws addressing development in the WUI. But they vary widely; and most are insufficient. Arizona and New Mexico, for instance, reject state mandates and leave it to cities and counties to decide how the WUI should be developed.[62] As you might imagine, that doesn't limit risk very much. A fire ripping through a less protected county doesn't skid to a stop when it nears the border of a more protected one. And with local governments competing against one another for development, council members find themselves racing toward the least-restrictive common denominator. In contrast, California and Oregon have each adopted *statewide* risk-reduction standards for wildfire, specifying the types of architecture and building materials that are permissible in the WUI.[63]

But even states with uniform standards could go further. Just as communities impose tougher requirements in federally designated flood plains, states could restrict uses and building methods in districts or neighborhoods that are especially prone to wildfire. Of course, that would first require fire-hazard maps calibrated to very local conditions, which the Forest Service's maps are not. But maybe, given proper funding and authority, the Forest Service could build in more precision and identify districts of special concern, just as FEMA designates floodplains for the National Flood Insurance Program.

The problem is that unless such maps are paired with a plan to encourage affordable housing development in safer places, then this would just lead to higher home prices and more inequality. We need incentives for infill development in larger cities, particularly near zones of economic development where jobs are. One solution might be "urban growth boundaries." These are enforceable limits that separate how far an urban area can expand into the rural landscape. Portland, Oregon, and Knoxville, Tennessee, both have them. Los Angeles is considering the approach.[64] The key, though, is to make sure there is enough room, using infill and multistory apartment buildings, to provide decent housing for those living within the chosen boundary. This is where cities like Portland have historically failed.

Another tough option involves retreat. Some places may be so hazardous that residents should be encouraged—maybe required—to relocate. In chapter 6, we looked at retreat in the context of flooding. (We'll see it again in chapter 11.) Retreat from the line of fire might be accomplished through voluntary buyout programs, though anything at true scale would probably be prohibitively expensive. In the meantime, individual owners drive the market one decision at a time. The town of Paradise is building back, with thousands of building permits in line for processing. Dozens of burned-out communities in the Cascade Range of Oregon and Washington are also rising up from the ashes.

Yana Valachovic, a forest adviser for the University of California Cooperative Extension in Humboldt and Del Norte Counties, understands that people's choices are limited. "What if all you have is insurance money to rebuild [on the same lot] or sell the property way below its value to someone else? How would you make that choice?" Valachovic asked. "And if people choose to stay, who is going to issue an eminent domain proclamation and take private property, and does that make sense fundamentally? The entire American West is wrestling with these questions." In the end, she says, a lot of people just want to rebuild. "There's immense pressure on local politicians to respond to personal demand and that's true in flooded areas and hurricane zones just like it is with fires."[65]

The housing challenge is complicated, but like forest management, society has the technology and the wealth to make big improvements if we can shift the culture. Sam Perkins fights wildland fires for the Forest Service, but he has also taken time off to sharpen his skills as an adviser and advocate. After he graduated from Whitman College in Walla Walla, Washington, Perkins took a fellowship that allowed him to study resilience and natural disaster in countries like Japan and Nepal. Years later, he was hired by Nepal's ministry of forests to help develop that country's first wildfire program. Cinthia Zermeño Moore is shifting the culture too, through her work with Moms Clean Air Force and other organizations. She is educating her community about public health issues, getting involved in local politics, and inspiring others on radio shows, podcasts, and webinars to do the same.

Want to fight wildfire as part of your climate resilience work? Learn about the state or national forests in your vicinity. Go online and look at their management plans. When those plans come up for review, take advantage of the public participation process and insert yourself in the decision making. You don't have to do this on your own; there are local and national advocacy groups that would love to have your help.[66] Or wade into the WUI. Check out the Forest Service's "Wildfire Risk" mapping tool and learn about the wildfire hazards in communities you care about. Find out how residents and local leaders are addressing those risks. Educate people and push for change. Maybe you live outside the WUI in Seattle or Sacramento or Sarasota. Your city's housing policies could be pushing families out of the safe districts and toward less populated places that are more prone to burn. Support organizations and businesses that are helping to expand affordable housing options in your city.

Against the apocalyptic boom of today's fire seasons, I can't blame you for wanting to cover your ears and bolt. But if there's anything I've learned after years of kayaking Useless Bay, it's that you can't control the boat without putting your paddle in the water. And when the blackberries are sweet, a little dust doesn't matter.

chapter 9
Yuccas, Gardeners, and Zookeepers

Yucca *brevifolia*, commonly known as the Joshua tree, casts a cartoonish shadow when the sun slides behind the mountains. The plant's looping arms and bayonet leaves already cut a strange profile. But at dusk, its silhouette undulates across the desert like something out of Dr. Seuss.[1] It can make you giggle. And as with a Seuss book, the flutter it sends to your heart can feel both blissful and menacing at the same time. Or so it seemed to me that March evening in the backcountry of Joshua Tree National Park.

A couple miles from the marked trail, I had already descended a steep slope and found a promising spot for my tent, just in the shadow of a multiple-limbed Joshua. It was a healthy specimen, about twenty feet high. From the tips of its spiked branches drooped large artichoke-like blossoms, lush and creamy with white petals. It was perfect. I wanted a site where I could feel a part of the landscape, surrounded by Joshua tree, creosote bush, prickly pear, and all the rest. The last site I had found wouldn't do. It had been recently hit by fire, and there were not many Joshua trees left standing. Most had been reduced to stumps, their insides crumbling like baked Styrofoam. One of the bigger ones, shriveled and prostrate, reminded me of a fallen soldier.

At the new site, there was more going on—more *life*. A cloud of birds swirled in the air. An owl screeched. A jackrabbit vaulted across a gulley and into a creosote bush. I spent a surprising amount of time watching a pocket mouse rooting for seeds in a layer of crusty sand. The wind picked up, and gusts of cold air poured into the valley. When the last sliver of sun dropped, the shadows receded, and I searched my pack for a headlamp and my down jacket.

Having been born and raised in Las Vegas, I grew up around Joshua trees. As a kid, I hiked among them in Red Rock Canyon with my grandfather. At home,

I studied their silhouettes as printed on the backs of playing cards that my grandmother brought home from the Desert Inn Hotel and Casino, where she worked. I even have faint recollections of visiting the Joshua Tree area back when it was still a national monument. But I had never walked the landscape as a national park property, and I had never camped in the backcountry.

This time I was visiting Joshua Tree to get some answers about climate change. Our federally owned lands—which provide economic resources, opportunities for spiritual and recreational renewal, and habitat for thousands of species—are unraveling as the climate breaks down. Slowing this decline is imperative, but even the experts don't know how to do it. One theme in this book is that we'll never save everything. Building climate resilience means setting priorities, deciding what battles to take on and what efforts to let go. It means gauging the breeze, eyeing the currents, and reckoning how safely you can get from A to B. Field biologists and climate scientists are producing better data every day. But the question of priorities and effort is still basically a political and legal one. The best biologist's smartest idea will never launch if there's no wind in the canvas and no sailor at the helm.

But back to the desert. The Joshua tree has become a poster child for the climate crisis, especially in the American Southwest. For reasons we'll see, the iconic plant is in peril and could disappear completely from Joshua Tree National Park under some climate scenarios. The tourist industry has taken note. So just as campers are urged to see the glaciers of Glacier National Park while they still can, publications now urge travelers to see *Yucca brevifolia* before it lands in intensive care. Or as a headline in the *Los Angeles Times* put it: "See Those Joshua Trees While You Can—Climate Change Is Killing Our National Parks."[2]

By visiting the park and learning about a particular climate challenge on our federal lands, I thought I might gain insight into the problem as a whole. In fact when I contacted park officials about my plans, they suggested I first visit the elevated backcountry, a wilderness area especially conducive to the Joshua tree. As average temperatures climb and conditions become less hospitable, areas like these are thought to be the best last chance for Joshua trees in the park. Ecologists call these special places "refugia." Of course, the fire damage I saw suggests that not everything is fine, not even there. But on that spring night with a cloudless sky and a swirling wind, the temperature predictably dipped below freezing, just perfect for these plants, and the stars that night were as exquisite as any I've seen.

To get some details out of the way, the Joshua tree is not actually a "tree"—or a cactus, either.[3] It's a yucca, which explains the pointy leaves and summertime blossoms (figure 9.1). And it's a *very big* yucca. In the park's backcountry, I estimated some to be at least thirty-five feet tall. In the suburban neighborhoods of Las Vegas I've seen taller ones, but they're fed by sprinkler systems. No one is sure why Joshuas grow so big—there's an evolutionary cost to living large on a barren landscape. But some scientists think it might make them better able to withstand brushfires.

Unlike real trees, the pulp of Joshuas produce no annual rings, so one can't say for sure how old a Joshua is or how fast it grows. The best guess is that the plant's natural life span tops out at 150 years. It's thought that three-quarters of all living Joshua trees are less than a hundred years old. Observational studies show varying rates of Joshua tree growth, but everyone agrees that, at least in current times, the rate is astonishingly slow. One study over a twenty-eight-year

9.1 *Yucca brevifolia*, Joshua Tree National Park.

(Photo by Rob Verchick)

period found that Joshua trees in the national park had grown an average of 0.13 inches per year.

The Joshua's singular height and soft wood make it a prized housing option for small desert dwellers. Red-tailed hawks nest in the tree's upper branches. Doves and cactus wrens squeeze themselves between spiky leaf rosettes. Wood-peckers and northern flickers drill nest cavities into dead trunks. When they vacate after a year, their hollows give secondhand housing to western bluebirds, house finches, fly catchers, screech owls, and more. With all the birds—and their eggs—it is not surprising that foxes, bull snakes, and other predators take up residence nearby. Joshua trees, like other succulents, also serve as vital watering stations in the desert. Pocket mice, wood rats, antelope squirrels, and jack rabbits have all been found tapping free water stored in the yucca's thick stalk. Across broad swaths of the Mojave, no other plant can fill this role in all four seasons.

Among the human population, the giant yucca has been both loved and loathed. Native peoples admired the plant for its durable fiber and sustaining buds and seeds. Mormon pioneers emigrating from Salt Lake City to Southern California in the late 1800s would have recognized the yucca as a happy sign that their journey had crossed the midpoint. The tree's outstretched limbs, it is said, reminded them of the biblical Joshua, beckoning his people toward the Promised Land. But observers have also described it as "tormented," "unhappy," and "the most repulsive tree in the vegetable kingdom."[4] "One can scarcely find a term of ugliness," wrote the naturalist J. Smeaton Chase near the turn of the last century, "that is not apt for this plant." But by the 1970s, the desert writer Peggy Larson had crowned Joshua to be "the Mojave's most distinguished plant," and today's travel guides routinely rhapsodize about its gnarled beauty and enduring spirit.[5]

About fifty miles east of Palm Springs, Joshua Tree National Park lies at the convergence of two southwestern deserts, the higher Mojave Desert and the lower Colorado Desert. Elevations range from about five hundred feet to over five thousand feet. Outside the boundaries, literally across the street from the northern entrances, is the Marine Corps Air Ground Combat Center. That facility, the largest U.S. military base in the world, is currently used to train soldiers before deployment to the Middle East.

Located at the southernmost limit of areas where Joshua trees grow, the park is said to provide habitat for over 800 higher plant species, 250 bird species, and more than 100 species of mammals and reptiles. Over the last ten thousand years, the area has hosted several permanent human settlements too, including those of the Pinto Culture, the Serrano, the Cahuilla, and the Chemehuevi.

Today, the region is dominated by outdoor tourism and the military. Inside the park boundaries you'll find hiking, birdwatching, and perfect viewpoints for amateur astronomers. With such easy access to gigantic boulders and rock outcroppings, Joshua Tree has also become a magnet for day-tripping rock climbers. "Belay from your bumper" is actually a saying around here. (But please don't—it's dangerous.) Indeed, if you've not been to Joshua Tree, you've probably already seen glimpses of the landscape in movies or on at least one celebrated album cover, Eagles' 1972 debut record, whose cover featured a glowing raptor soaring above the arms of a yucca. (The cover shot for U2's *The Joshua Tree*, its name notwithstanding, was taken at Zabriskie Point in Death Valley.)

For all this, we can thank Minerva Hoyt, whose activism over three decades won the area protected status. Born during the Civil War to a wealthy Mississippi family, Hoyt spent her early adult years in New York and Baltimore before settling in Pasadena, California, in 1897. She and her husband, a wealthy physician, were quickly absorbed into the city's social elite. She got involved in several civic organizations, including the prestigious Garden Club of America. In those early days, Hoyt gained deep affection for the range of California's desert flora. But she fell especially hard for *Yucca brevifolia*. At the same time, the expansion of roads into the state's desert was pulling in not just nature buffs, but land developers and cactus poachers, as well. Hoyt grew alarmed.[6]

"Over thirty years ago, I spent my first night in the Mojave desert of California," she recalled in 1929, "and was entranced by the magnificence of the Joshua grove in which we were camping and which was thickly sown with desert juniper and many rare forms of desert plant life. A month ago . . . I visited the same spot again. Imagine the surprise and the shock of finding a barren acreage with scarcely a Joshua tree left standing and the whole face of the landscape a desolate waste, denuded of its growth for commercialism."[7]

Hoyt launched a personal mission to save the remaining landscape that remained. To call attention to her cause, she staged lavish exhibitions of desert plants in cities like Boston, New York, and London. Organizers would receive freight cars full of crusty rocks, sand, cacti, yucca—even a stuffed coyote—and pose them in elaborate displays. Desert flowers were flown in twice a week and stored in Hoyt's own hotel bathtub before installation. The shows were wildly successful. By the early 1930s, Hoyt had built an international reputation for desert conservation and was lobbying hard for federal protection of the Mojave's landscape. To cap the effort, she commissioned an album of glorious landscape photographs by Palm Springs artist Stephen Willard and managed to have it personally delivered to President Franklin Roosevelt in the summer of 1934. The president was impressed. Two years later, he signed a presidential proclamation establishing Joshua Tree National Monument, a preserve of more than 800,000 acres dedicated to the protection of "various objects of historical and scientific interest"— legalese for Hoyt's "magnificent" and "thickly sown" arid plant life.

In 1994, Congress upgraded the monument to national park status and reconfigured its boundaries to include whole ecological units, including the Little San Bernardino mountain range. Of the current 792,623 acres, three quarters of them are designated "wilderness," a classification affording the law's strongest protection. In upgrading Joshua Tree's status, Congress didn't mention climate change by name, but it noted that California's desert resources "are increasingly threatened by adverse pressures" that could one day destroy them. Among other things, it directed the Park Service to "retain and enhance opportunities for scientific research in undisturbed ecosystems" and "perpetuate in their natural state significant and diverse ecosystems of the California desert."[8]

———◆———

Upon my return from the park's backcountry, I had arranged to meet two park scientists at their offices near the Oasis Visitor Center. A Mediterranean-style building of wood and stucco, the center is surrounded by specimens of fan palms, beavertail cactus, and big berry Manzanita. On the front exterior wall there's a mural in which a life-size Minerva Hoyt stands serenely in a wide-brimmed hat and full-length dress, the arms of a blooming Joshua tree gesturing behind her. Neil Frakes, who manages vegetation programs at the park, met me at the mural. He's a bearish man with horn-rimmed glasses and a fuzzy beard. Frakes gave me a tour of the complex, which included a peek into a greenhouse containing specimens of desert plants and green rows of Joshua saplings.

Inside the office, Frakes introduced me to Jane Rodgers, the park's chief scientist. Rodgers is known for her deep expertise in restorative management and desert ecology. She fell in love with habitat conservation in the 1990s as a Peace Corps volunteer in Niger and has spent her career working on large-scale restoration projects.

Most of what occupies Rodgers's and Frakes's times is in one way or another driven by the climate crisis. Simply put, the southern California desert—which since my childhood I have associated with an enormous convection oven—is getting too hot. Droughts are becoming harsher and more frequent, and that is affecting the home ranges of some plants and animals in the park.

"We're already seeing less water in the springs for the bighorn sheep," Rodgers told me. Rainfall is critical to the health and reproductive success of desert species. Severe droughts can cause desert tortoise and bighorn sheep to migrate to higher elevations that receive more rainfall. To help bighorn sheep in the summer, Rodgers explained, artificial catch-basins called "guzzlers" have already been installed outside the park to supplement natural watering holes. This sort of light "zookeeping," I've learned, is not unusual in and around public lands, particularly in the arid West.

Some experts have suggested artificial irrigation might be necessary to support Joshua trees. But Rodgers sees that as unlikely.[9] Instead, park officials are focused on identifying refugia, like the mountains I camped in the night before, where Joshua trees are thought to have the best chance of surviving in a warmer climate. There is more rainfall there; and, as I learned, the mazelike topography has a way of forcing chilly winds from the granite peaks into the sunny valleys. The plan is for park officials to protect those landscapes as best they can, in hopes of preserving something of the original yucca-dominated landscape. There are challenges. The refugia make up at best 19 percent of the plant's current park habitat, and experts are not sure whether long-term survival is possible even in those cooler pockets.[10] It's not just a matter of having enough water. A whole clan of interdependent species have to be able to live there for the habitat to function properly. Some biologists, for instance, fear that a certain pollinating moth important to the Joshua tree's reproduction cycle may be declining and threatening the plant's survival throughout the park.[11]

But the most immediate threat, according to Frakes, is wildfire. He pointed to a topographical map pinned to a very messy bulletin board. The map was in grayscale except for wide bands of red and orange—indications of recent fire—fanning across the park's northwestern quadrant.

Lightning fires have always been part of the Mojave's ecological cycle. But there was limited damage because shrubs and trees were spaced widely apart, and connecting grasses were sparse. Before 1965, most lightning fires burned less than a quarter of an acre. Since then, the number and intensity of fires in the region have exploded. In 1979 a fire on Quail Mountain, the park's highest peak, burned six thousand acres. In 1995, a fire in Covington Flats burned more than five thousand acres. In 1999, the Juniper Complex fire—the largest in Joshua Tree history—destroyed nearly fourteen thousand acres of California junipers, pinyon pines, and Joshua trees.

Without much protection beyond their spiny leaves, Joshua trees ignite easily and burn quickly. Not all die. Even where a landscape is leveled, some percentage of trees will usually survive and generate new sprouts the next season. But the growth rate is notoriously slow, and it can take decades for a mature grove to reestablish itself.

The trend toward bigger and more frequent fires now poses an existential threat to many Joshua populations in the park. When Frakes pulled the map from the board to give me a closer look, I could see why. Also included on the map were small black patches indicating areas of Joshua tree refugia as projected under a 3 degrees C warming scenario. Most were huddled in areas of higher elevation, with names like Black Rock Canyon, Sheep's Pass, and Lost Horse Valley; many of those places were in the damage zone. Referring to the refugia, someone had scrawled in black marker across the bottom of the map: "1/3 Burned w/ 10–15% Survival."

I asked Frakes what could explain such a dramatic expansion of desert fires in such a relatively brief period of time. It didn't seem to me that incremental changes in climate could adequately account for that.

"Cheatgrass," he said, rolling his eyes. "And red brome. We have a hard time controlling them." Cheatgrass and red brome are two related invasive grasses that are fanning across the Mojave, Colorado, and Great Basin deserts at record speed. Fast-growing annuals, they fill in the gravelly gaps between shrubs and yuccas. When they dry out in the summer, they become a potent source of fuel, enabling fires to roar up mountains and tear through valleys. Although many desert plants can resprout after fire, the process is slow. It can take a small shrub like black brush fifty years to return to its pre-fire density. Cheatgrass and red brome, on the other hand, are better adapted. It is a wicked irony that the same fires that torch the grasses in summer also help expand their range by clearing out the native plants that would otherwise compete against them in the spring.

Rodgers wanted to make clear that the invasion of fire-spreading grasses in the American Southwest is not directly caused by climate change. Nevertheless, climate change is making severe wildfire conditions more common. The stress that higher temperatures puts on desert species, including Joshua trees, makes them less able to recover when fire strikes and, over time, weaker. The result is a bundle of feedback loops that scientists still do not fully understand.

Until scientists have a better understanding of these relationships, park policy now requires the suppression of all fires of any origin within Joshua Tree, including those in wilderness areas. In addition, Frakes is overseeing an ambitious program of "fuel reduction" in which crews of workers are brought into certain parts of the park with motorized equipment to mow down invasive grass and prune away woody shrubs, including native ones. Park officials are also considering the limited use of herbicides. After a fire, park crews work to restore certain areas by planting Joshua saplings taken from the greenhouse Frakes had shown me earlier.

All of this seems relatively new at Joshua Tree. The fuel reduction crews, for instance, work on contract. "It's the Forest Service that has the expertise in thinning," Frakes told me. And committed funding streams are embryonic. At least some of the fuel-reduction work is drawn from a fund to protect roads and infrastructure. It should be stressed that neither the thinning nor the replanting has ever taken place in the more protected wilderness areas, where many of the refugia are located. I asked if restoration of any kind took place in wilderness areas damaged by fire.

"Never really done it," Frakes said. "We've thought about it."

"We take pride in letting nature be nature," Rodgers added. "We have a culture of not meddling."

But, of course, the meddling has already begun. Indeed, climate change is disrupting the entire park system at an unprecedented level. We know because around 2016, the director of the National Park Service, Jonathan Jarvis, finally began saying so to nearly anyone who would listen. A Park Service "lifer," Jarvis began his career forty years ago as a seasonal ranger in Washington, DC. He headed several park units in the West and ultimately served as Park Service director from 2009 to 2017. Managing more than 400 parks and 22,000 employees on a limited $3 billion budget, Jarvis faced a stampede of challenges. But as the Park Service moved into its second century, what kept him up at night was what he called "the c-word."[12] Climate change, he said, was "the greatest threat to the integrity of our national parks that we have ever experienced."[13] Jarvis warned about climate change in forums all over the country, including a TED Talk. Enacting bipartisan legislation aimed at reducing the nearly $20 billion maintenance backlog in national parks and on public lands would cost $6.4 billion over the 2019–29 period, according to an estimate from the U.S. Congressional Budget Office.[14]

Many of Jarvis's talking points relied on a remarkable study published in 2014 by Park Service ecologists William Monahan and Nicholas Fisichelli.[15] Using Park Service data, they surveyed nearly three hundred parks from the Pacific Islands to the eastern seaboard. They examined each park with reference to twenty-five variables relevant to climate change, from "annual mean temperature" to "precipitation of the driest month" to "mean annual percentage cloud cover." The goal was to see how current conditions at each park compared to its "historic range of variability," a span defined from 1901 to 2012. Monahan and Fisichelli found that the parks were "overwhelmingly at the extreme warm end of historical temperature distributions," but showed much greater variability in precipitation patterns. By historical standards, some parks, like Kalaupapa National Historical Park in Hawai'i or Lake Mead National Recreation Area in Nevada and Arizona, were warm and withered. Others, like North Carolina's Lookout National Seashore or Colorado's Florissant Fossil Beds National Monument, were heated and soaked.

Some patterns developed. Parks in the southwestern deserts and in Hawai'i were warmer and drier; parks in the Northeast were warmer and wetter; and parks in the Midwest were warmer but showed little change in precipitation. In the Southeast, the researchers found that temperatures were rising *more slowly* than in neighboring regions, an effect consistent with what scientists call the southeastern "warming hole"—a void where the continent's wider trends seem somehow suspended in time. (The cause is debated but may be related to exhaust from coal plants.)[16] At least one park, Apostle Islands National Lakeshore in Wisconsin, had already reached the top of its historic range for temperature variability, which is to say that conditions at the "Jewel of Lake Superior" will soon be unlike what any living creature has ever seen there. As trends in warming and precipitation continue, Monahan and Fisichelli expect many more parks to cross into this realm.

These conditions will have many effects. Yet we can imagine some of what the new version of our park system will look like, because, as we've seen with Joshua

Tree, it's already changing. Montana's Glacier National Park lies in a mountain range that a century ago held more than a hundred glaciers. Now all but twenty-five have melted. Scientists say that in less than ten years there may be *no* glaciers in Glacier National Park. (Glaciers are also retreating in Alaska's Glacier Bay National Park and in Washington's Mt. Rainier National Park.) In other regions, rising seas are the major threat. The Park Service estimates it has billions of dollars in infrastructure within three feet of current sea level, which includes the shoreline around Ellis Island and the Statue of Liberty.[17]

Still, there is more to our public lands than national parks. In all, the federal government owns about 2.27 billion acres, or 28 percent of the nation's landscape.[18] In addition to national parks, there are federally held wilderness areas, wildlife refuges, forests, grazing lands, Indian reservations, and more. These lands often are managed by different agencies (or, in the case of reservations, by native tribes) with sometimes very different procedures and goals. Nonetheless, all of it has shaped our history, our traditions, and a basic sense of who we are. Federal lands also stoke the economy. From Nevada's sagebrush steppes to Florida's rivers of grass, federally managed ecosystems deliver tens of billions of dollars in goods and services—things like timber, water, food, medicine, fuel, flood protection, fire suppression, biodiversity, and a booming market for outdoor recreation.

But, as in the National Park System, climate change is rejiggering the system.[19] In the Sierra Nevada and the Intermountain West, temperature fluctuations are upsetting the natural plumbing. Declining snowpack in central California impairs the quality and quantity of water that feeds important rivers.[20] In the southern Rockies, shifts in the timing of runoff are linked to floods caused by overtopped dams and reservoirs. Droughts, of course, also threaten water supply and in the Southwest have contributed to lower stream flows, more intense wildfires, and the disappearance of several lakes.

In our national forests, changes in temperature and precipitation foretell a plague of pests, pathogens, and drought stress. Scientists, for instance, blame global warming for an outbreak of mountain pine beetle that killed hundreds of thousands of acres of trees across the American West. The beetle and western pines have coexisted for thousands of years, but recently warmer winters seem to be giving this quarter-inch bug the upper hand. Able to survive the winters more easily, their numbers have grown exponentially, and they have vastly extended their range. Beetle attacks, coupled with the introduction of a pathogen called blister rust, have caused a massive loss of one particular tree—the whitebark pine—a "keystone species" that many animals, from Clark's nutcrackers to grizzly bears, rely on for food. Finally, climate disruption is opening the gates to an array of nonnative plants and animals that, like the pine beetle are finding some habitats much cozier than before. Impacts like these are threatening every corner of our shared landscapes. From the redwoods to the Gulf stream, Woody Guthrie sang, "this land was made *for* you and me." Someday our grandchildren will look back and realize that This Land was made *by* us, too.

<hr />

Scientists have known for a long time that climate breakdown threatens the Southwest. But it was only about fifteen years ago that Joshua trees became a lens through which the climate story could be viewed. For that, we can credit Kenneth Cole, a climate scientist at the Colorado Plateau Research Station in Flagstaff, Arizona. A study he published with colleagues in 2011 slapped a climate bull's-eye on *Yucca brevifolia* and inspired a host of later investigations into the vulnerability of southwestern landscapes to climate change.[21] It also boosted the public's fascination with—and concern for—Joshua Tree National Park. It was Cole's paper, Frakes told me, that "put the foot on the gas."

Cole set out to answer a simple question: given climate change and absent human intervention, where would Joshua trees be living at the end of this century? Understanding his work helps us grasp the interdisciplinary nature of climate change questions and the challenge of tracking species into the Anthropocene's misty future. To forecast the future location of Joshua trees, Cole set out to build a cartographer's version of a time machine. He needed a mapping tool that would allow him to view yucca colonies in the deep past, compare them to the present, and then draw conclusions about the tree's prospects in the near future. The work involved traditional mapping, high-resolution computer modeling, and the fossilized dungballs of giant sloths.

Cole started with the present. Culling through federal and state databases of known Joshua tree populations, Cole's team constructed a gridded map displaying the plant's presence across the American Southwest in the mid-twentieth century. That map showed what most people familiar with the Southwest would already expect—plumes of *Yucca brevifolia* radiating from Joshua Tree National Park, the Mojave National Preserve (seventy miles northward) and Red Rock

Canyon. By comparing the temperature and precipitation of occupied areas with unoccupied areas in the same region, Cole's team was able to speculate about the kind of climate conditions necessary for Joshua trees to thrive.

Would those climate conditions hold through the remainder of the century? To find out, Cole's team next consulted several widely accepted computer models used to forecast climatic change around the world. These models are reliable, but their resolution is "coarse": a typical grid cell on a map using such models is 250 square kilometers, or a little less than 100 square miles. Joshua Tree National Park is only 1,000 square miles, so that's not a very refined picture. To get a better look, Cole's team had to "downscale" the traditional models by combining observational data about western North America with global models. The downscaled models suggested that the climate conditions supporting today's Joshua trees in the Southwest would not hold. Specifically, they projected "a severe decline in the area of suitable climates for Joshua tree by 2070 to 2099 AD, perhaps to as little as 10 percent of its current range."[22] On the positive side, the downscaled models projected that a few areas not currently hospitable to the yucca would become so by the end of the century. Squint at the map and you could imagine new stands of Joshua tree thriving in Utah's sandstone canyons or Nevada's mountainous Great Basin Desert.

Were Joshua tree populations capable of moving to places like that? Many plant species, like many animal species, are capable of migration. Joshua trees, for instance, have this ingenious way of cloning themselves by growing lateral roots that creep underground and then sprout up into new trees. But because new clones sprout only a few inches from the original, that takes a long time. In addition, many desert plants are known to migrate through the digestive tracts of mammals and birds: a hungry jackrabbit, say, might devour a yucca's seed pod, scamper across the desert, and distribute the undigested pips in its droppings. Maybe desert herbivores offered an express lane to Joshua tree survival.

To track Joshua tree populations while on the move, Cole and his team would have to travel back in time nearly twelve thousand years ago, toward the beginning of the Holocene epoch. The Holocene is the last time in which the average summer temperatures in North America were as warm as they are today. (The effect was prompted by a slow shift in the tilt of Earth's axis, a well-understood phenomenon that has no relationship to the climate change we see today.)[23] Fossil records show that before the warming period set in, large stands of Joshua tree extended from today's Los Angeles County to Phoenix and as far south as

Baja California. But about twelve thousand years ago, the summers warmed dramatically and, subsequently, Joshua tree populations sharply contracted. They survive only in the northernmost regions, mostly where we see them today. While there were new habitats opening up in the north that might have supported the embattled yucca, the populations migrated too slowly to catch up with them. Fossil records suggest Joshua tree colonies migrated less than six feet per year. That is much slower than the rate of many surrounding plants. It seems unlikely that *Yucca brevifolia* could have conquered the pre-Holocene Southwest at such a leisurely pace. Thus two questions come to mind. How did Joshua trees manage to cover so much territory in the pre-Holocene? And why, only a few thousand years later, did they stop expanding?

This is where the giant sloth comes in. While Joshua trees were expanding their range in the southwestern deserts, a mammal called the Shasta ground sloth was enjoying an equally successful historical moment. Today we think of sloths as confined to the rainforests of Central and South America. But in the Pleistocene epoch, which preceded the Holocene, the Shasta ground sloth roamed widely on lion-colored savannas that today make up the deserts of Arizona, New Mexico, and West Texas. Amid scattered clumps of juniper and prickly pear, they shared the landscape with mammoths, camels, and packs of howling dire wolves. Like other top herbivores of the Pleistocene, ground sloths were enormous, weighing around four hundred pounds and measuring eight feet from snout to tail. That made them the perfect match for the towering, lance-leaved Joshua tree. Back then a hungry sloth would rise on its hind legs, lean against the plant's thick trunk, and rip its fibrous arms to shreds, slurping up bushels of avocado-sized seed pods in the process. Thus satisfied, Shasta sloths would distribute Joshua tree seeds widely across the grasslands in the form of dungballs, some now fossilized and available for scientific study.

It is through the study of those dungballs that Kenneth Cole and his colleagues were able to identify the range of Joshua tree habitat in the Pleistocene and to speculate on the sloth's vital role in seed distribution. We don't know for sure how the Shasta ground sloth went extinct, but we have a strong hint. The demise of the sloth corresponds well with the appearance of human beings—nomadic tribes of Paleo-Americans, to be exact, who came armed with spears tipped with exquisitely sharpened stones. Thousands of years later when summer temperatures jumped, the Joshua tree's range shriveled up. Even after the climate again cooled, the population never rebounded. Without sloths, the

responsibilities of seed distribution had devolved to rabbits and ground squirrels which, try as they might, had neither the mass nor the range to launch a new
migratory mission.

Embedded in the yucca's long history is the insight that anthropogenic climate
change may not mark the first time that human activity has affected the Joshua
tree and the expansive ecology it is bound up in. Score one for the sloth hunters.
But what a difference a few tens of thousands of years can make. Now, we can
understand and measure the effects our activities are having on our landscapes
and around the world. And we have the moral and social structures—along with
the political and economic mechanisms—to evaluate those activities and change
course.

In the 1970s, Edward Abbey, the former park ranger, essayist, and desert cur-
mudgeon, complained that even the natural erosion of rock formations in
Arches National Park was once up for debate. "There have been some, even in
the Park Service," he wrote in his book *Desert Solitaire*, "who advocate spraying
Delicate Arch with a fixative of some sort—Elmer's Glue perhaps or Lady Clairol
Spray-Net."[24]

Abbey's lament sounds over the top, but the basic story has since been cor-
roborated.[25] It raises two questions that have nagged land managers since the
early days of the republic. How should this land be *used*? And what are we willing to *do* to make that so? In this case, those on the Spray-Net side wanted to
harness the land for its iconic scenery and were prepared to freeze the course of
natural erosion to do it. Abbey, on the other hand, liked knocking about the
desert in its naturally dynamic state (he was using it, too), even if it meant one
day losing something beautiful that he and others so obviously treasured.

The idea of stable baselines has always been a fiction. We might long for an
eternal landscape, but that's not how nature works. We might similarly long for
a landscape untouched by modern civilization, but that's not how humans work.
There is simply no place on the planet—not the South Pole or the Marianas
Trench—where you can't find industrial chemicals or shrimp gnawing on plastic
candy wrappers.[26]

In the 1970s, Canadian ecologist "Buzz" Holling pioneered a theory now
called adaptive management.[27] The idea was that because ecosystems are

dynamic, as well as subject to pervasive human activity, they require *mainte-nance*. He thus recommended policies focused on collecting information, setting measures of success, monitoring outcomes, and altering course when things didn't go as planned. It was a lot like gardening. And, like gardening, the process required *choosing* what you wanted to achieve, knowing it was impossible to have it all. If anyone needed convincing at the time, climate change tells us in the most compelling terms that no natural system will continue exactly as it has and that the nature our grandchildren grow to love (or loathe or fear) will be the nature we help tend.

But how far should we be willing to go? Should the National Park Service install sprinkler systems to save California's giant sequoia or let them fade into the mist with mild palliative care? What about allowing logging in sensitive pine forests to prevent beetle infestations? Would you airlift a community of mountain-dwelling pika from the Sierra Nevada to the Cascades in search of better climes? Or truck a bed of Joshua saplings to the slickrock of southern Utah?

Today's managers are trying an array of strategies, some with roots in tradi-tional conservation strategies, others evoking the "Hail Mary pass." They are try-ing to reduce strain on already stressed-out species. They are clearing wildlife corridors for species in search of better futures. In more extreme cases, they are implementing triage systems to identify the species with the greatest chance of survival and pondering heroic relocation plans. Sometimes, after careful thought, they choose to do nothing.

Most of the adaptation work these days is about buying time. That means shielding ecosystems from climate-change impacts for some period of time while we work on reducing carbon emissions or reengineering the climate. The fire prevention strategies at Joshua Tree National Park are examples of this. But in other places, fire builds resilience and helps buy time. For instance, in Washing-ton's Olympic National Forest, current stands are probably too dense to flourish in a warmer climate of drier soils and more pests. So park staff are using con-trolled burns to thin the forest and help surviving trees bulk up faster.[28] As the wildlands surrounding Minnesota's Boundary Waters warm up, biologists fear that an invasion of heat-seeking jewel beetles could devour important stands of black ash. The Forest Service, they say, might consider a nearly unprecedented use of pesticides to protect them.[29]

In the long run, though, buying time isn't enough. More permanent survival will require reshuffling the ecological deck, what biologists call "realignment."

That means making sure that restored or relocated habitats have all the ingredients necessary to keep an ecological system humming, from soil nutrients to pollinators to apex predators. Nature, of course, is a master at realignment. But global warming and mass extinctions are changing the planet too fast for nature to keep up. So, scientists and land managers have cautiously started experimenting. In southern Colorado, for instance, the Forest Service and the Bureau of Land Management (BLM) are using climate projections to help land managers plant different tree species that are better adapted to fire, drought, and pests. In one grassland revegetation project at the base of Colorado's Front Range, managers revegetated a gravel pit with an "uncertain mix" of native grasses representing a range of different moisture demands.[30] The tallgrass community that emerged is different from what would have naturally sprouted; but it is thought to be capable of surviving invasions by nonnative plants and a multiyear drought. The trouble with this kind of reshuffling is that the ace you think you're laying on the table could really be a joker. Scientists are still uncertain about what should go where.

David Graber, who served as Park Service scientist for nearly forty years, once put it this way: "I've had a number of conversations with land managers, identifying all the land in California that could conceivably be used as refugia, and what would be the appropriate species to go where. The magnitude of the problem is mind-boggling." While Graber favors realignment efforts, he acknowledged that some of his colleagues do not. "There is a vocal minority of people in the conservation community who believe that things should unfold on their own," he said, "the theory being, we don't know what we're doing, and we're bound to screw things up."[31]

Then, even if you know you have the ace, it might not be enough. To introduce Joshua tree to southern Utah, for instance, you would need not only the potted saplings but also all the soil microorganisms, fungi, moths, birds, jack rabbits, and other things that support the habitat. And they would all have to adapt to that area on roughly the same schedule.

Accumulating the knowledge for successful realignment thus requires large investments in time and money. You need carefully designed monitoring programs and consistent funding over time. As federal officials have noted, there is a "history of unfunded monitoring programs and monitoring-related line items are often the first cut by decision-makers."[32] You also need agencies with the power and inclination to protect the science from political tampering.

Unfortunately, as you have probably guessed, the federal government is nowhere near making those kinds of commitments. But there may be a deeper policy problem confronting public land management in the era of climate change—something more fundamental than gaps in scientific knowledge and funding levels. According to a growing number of experts, that problem lies with Congress's own statutory framework, which tends to encourage active management in places we want to develop, but discourages it in places we want to preserve.

The tension between management and preservation shows up in the way we organize our agencies to steward federal land and in the laws that govern them. For instance, a main goal of the U.S. Forest Service is to manage forests for logging. The BLM has long focused on making public land available for grazing and mining. To facilitate those uses, Congress built lots of flexibility into the laws governing those agencies.[33] Each is charged with balancing "multiple" land uses so as to "conform to changing needs and conditions," while insuring a "sustained yield" of goods and services in perpetuity. If that sounds vague and idealistic, it's supposed to. In establishing these management agencies, lawmakers knew they didn't have the expertise or prescience to micromanage millions of acres. So they set the goalposts and gave the professionals free rein. According to one court, these "multiple use" standards "breathe discretion at every pore"—a memorable, if not biologically accurate, metaphor.[34]

At the other end of the spectrum are preservation-based standards, which we associate with national parks, wildlife refuges, and wilderness areas. While the National Park System began as a practical enterprise to advance tourism and public recreation, it has over time nurtured an ethic of ecological stewardship. The days of wolf eradication and staged bear feedings at garbage dumps are gone. Since the 1960s, the National Park Service has interpreted its congressional mandate to emphasize historical preservation and noninterference. If a management strategy might imperil a preexisting biotic community or ecological function, it won't happen. On the other hand, if a limited action might strengthen or restore a degraded bog or elk population, park managers have the discretion (and sometimes the duty) to make it so. In maintaining wildlife refuges, the U.S. Fish and Wildlife Service follows a somewhat similar course, using a soft touch to see that biological and ecological systems track "historic conditions" as much as possible.

Wilderness areas, like the Boundary Waters around Lake Superior or the yucca refugia in Joshua Tree National Park, work a lot like the overlay zones city planners use to protect historic districts. Congress can classify portions of federally owned land—including national parks, national forests, wildlife refuges, and BLM land—as "wilderness," an official designation established in the Wilderness Act of 1964.[35] When that happens, the agency charged with managing that tract must work to preserve its "primeval character." That means keeping human intervention to a minimum. Use of permanent structures, motor vehicles, or heavy machinery requires special permission and consultation up the chain of command. Wildland firefighters clearing debris after vast conflagrations have been known to attack their prey on foot with field hoes and handsaws.

I have a lot of sympathy for noninterventionist strategies like those used in national parks and wilderness areas. But preserving the past is not a recipe for winning the future. An emerging scientific consensus says that unless federal land managers get more aggressive in battling climate effects, they will oversee an environmental catastrophe of continental proportions.

Many land managers and legal scholars are now wondering whether the laws designed to protect our most treasured lands are instead dooming them. For instance, J. B. Ruhl, a legal expert who specializes in natural resources, predicts the logic of the Wilderness Act "will fall to pieces in the era of climate change."[36] William Tweed, a naturalist and thirty-year veteran of the Park Service, has repeatedly argued that the service's founding legislation, known as the "Organic Act," is incompatible with twenty-first-century management needs.[37]

I first started thinking about this during the first term of the Obama administration. A political appointee at the U.S. Environmental Protection Agency (EPA), I had been asked to serve on a White House task force charged with drafting a national strategy for adapting to climate change. Our work resulted in an executive order, issued by President Obama in 2013, that required federal agencies to study the effects of climate change and take actions to address them.[38] Contrary to what a lot of people think, presidents can't just order agency officials to do whatever they want. It is Congress that gives most agencies their powers and Congress that defines their methods and goals. If the president says to the Park Service, "Prepare for climate change *now*," the service can only do as much as its congressional authority allows. So when I served on the president's task force, we thought a lot about the things that federal agencies could and couldn't do. Our recommendations, like the president's executive order, were firm on the objective of preparation, but we had to leave room for them to decide how to accomplish that.

In response to President Obama's executive order, the land management agencies released a slew of studies, proposals, and rules. Though President Trump later rescinded the order, many of the policies still exist, giving us some insight into what land managers think their agencies have the authority to do.

It does not surprise me that the Forest Service now boasts "the most extensive adaptation planning and integration of adaptation into management processes of any of the federal land management agencies," according to legal experts Alejandro Camacho and Robert Glicksman.[39] Not only does the Forest Service have a history of strong climate leadership and a tight chain of command, but it also has a legal framework that encourages it. Congress, for instance, requires the Forest Service to keep an inventory of lands "reflecting changes in conditions and identifying new and emerging resources and values."[40] And in 1990 Congress specifically ordered the agency to examine how climate change is affecting forest resources. Forest managers now factor in climate change into much of their long-range planning and are even introducing nonnative seed stocks in some places where conditions are predicted to change significantly.

Adaptive work at the National Park Service is not nearly so ambitious. Park Service policy continues to lean away from active management unless it is to restore ecosystems already damaged by human activity; and even then, the purpose is to return the landscape to its previous "natural condition," rather than something that might thrive in the future. The policy, of course, follows Congress's preference for historic preservation as expressed in the Park Service's Organic Act, as well as in legislation pertaining to individual parks. Congress's purpose in creating Joshua Tree National Park, recall, was to "perpetuate" desert ecosystems "in their natural state."[41]

As for wilderness areas, there are virtually no activities in place to prepare them for climate change, regardless of whether they are located in national forests, national parks, or other units. The Wilderness Act, recall, restricts human intervention in order to keep conditions "as is." Promoting ecological health is not even a legislative goal.

Before we conclude that long-leash regulation is always better for resilience, I should note that the BLM's approach to climate adaptation is—there's no other way to say it—pathetic. As with the Forest Service, BLM's "multiple use" directive gives it considerable latitude to adjust its activities in response to changing conditions. But the BLM's strategy limits itself only to assessing those changes and offers little in the way of concrete adjustments. As of this writing, the agency has produced not a single management regulation that even mentions the words "climate change." The difference between the Forest Service and the BLM here probably has less to do with legal framework and more to do with professional culture. The Forest Service has undergone years of ecological soul-searching, and while there is still a long way to go, the agency's career staff can, in my view, be said to be taking its stewardship mission seriously. The BLM, in contrast, remains known for its TV-cowboy swagger and gluttonous appetite for oil and gas fields, strip mines, and fracking pits. It's hard to care about climate change when you're doing so much to enable it.

It's true, though, that the law's commitment to preservation in parks and wilderness areas poses a barrier to active resilience efforts. Camacho and Glicksman believe the laws governing national parks and wilderness areas should be amended to elevate long-term ecological health over historical preservation.[42] But not everyone in the policy and legal community is convinced. Eric Biber, a professor at the University of California, Berkeley, and Elisabeth Long Esposito, a forestry lawyer in Sacramento, are also recognized experts on the law of public lands. They argue that preservationist statutes are more flexible than most people give credit.[43] Land managers, they point out, have engaged in wolf reintroduction programs in wilderness areas for years, sometimes swooping into those places with court-approved helicopters and dart guns. And there is at least one example where wildlife managers in Montana assisted the migration of a type of cutthroat trout to parts of a wilderness area where they had never existed before.

Projects like these do come with an array of "procedural and substantive hoops."[44] That's good, say Biber and Esposito, because the law is helping to ensure that management plans are well designed and based on the very best science. It occurs to me that if park and wilderness managers are not somersaulting through hoops fast enough, it may be because managers' own conservationist tendencies, summed up earlier by David Graber, are getting in the way. They're afraid "we don't know what we're doing, and we're bound to screw things up."

When I asked Neil Frakes and Jane Rodgers about legal barriers to adaptation at Joshua Tree, neither seemed very concerned. Buying time through restoration and fuel reduction, Frakes reminded me, is already taking place in many parts of park's developed areas. He thought it was important to have less-managed areas in the wilderness with which to compare results. That's how science works.

Rodgers believed that the biggest barrier to resilience efforts was in fact not legal but policy-based. She told me that park priorities are heavily influenced by visitor surveys and that making Joshua Tree resilient to climate change has yet to show up as a main concern. This is despite the park's efforts to educate the public about climate change in educational materials, ranger-led talks, and exhibits at the visitor center.

A recent survey of federal land managers in the Mountain West supports Rodgers's point. Of the many hurdles to implementing adaptation projects, the ones most cited as posing a "big barrier" were budget constraints, insufficient staff, lack of perceived importance to the public, and lack of public demand to "take action." While it did not make the top of the list, the lack of a "legal mandate" was identified as a big barrier by nearly half of respondents. Another survey, published in 2010, found that employees at three national parks in Washington state overwhelmingly believed the Park Service's Organic Act thwarted progress on climate change adaptation.[45]

A few hours after my conversation with Rodgers and Frakes, I was again on my own, this time driving southeast through the park in an area called the Lower Covington Flats. The dirt roads had been washed out by a recent storm. As I swerved to and fro—going faster than I should have—mud splattered through my open windows and onto the upholstery of my rented SUV. On Frakes's recommendation, I was barreling down the Flats to see a different part of the park, outside of the wilderness area, where I was told I might find a contract crew cleaning up after a recent wildfire. (Frakes had suggested I spend my extra dollars to rent the four-wheel drive but had said nothing about keeping the windows up.)

You see roads and cars and all kinds of mechanized equipment on the lower flats. You will also find patches of privately owned property that were acquired before the land was federally protected. Such lots, surrounded by chain link fencing, usually contain a home of some sort—a stucco rambler, a travel trailer— and two or three highly excitable dogs. Inspired by my surroundings, I plugged my smartphone into the dashboard and hit my Eagles playlist. Then out wailed Glenn Frey, singing about running down the road trying to loosen his load.

The surrounding hills were barren of woody plants except for occasional patches of desert scrub and creosote bush. But cheatgrass —which Frakes had taught me to identify by its hairy, drooping leaves—was everywhere. It swayed and shimmered in the sunlight. Occasionally, I would see a fallen Joshua tree, shriveled and charred. Two days before, in the backcountry, a dead Joshua had reminded me of a fallen soldier. Here for some reason, they reminded me of those inflatable tube men you see advertising a sale at a tire shop, but with the air gone out. At one point, I saw I was being tailgated by something big and yellow, so I pulled over to let it pass. A convertible Hummer with a half-dozen passengers bouncing and shrieking in the back zoomed by. "*Adventure Tours*," it said on the side.

Back on the road, I turned a corner and came upon a line of packed lawn-and-leaf bags sitting along the shoulder, separated by piles of cut branches. In the distance, I heard the thrumming of gasoline-powered string trimmers. As I got closer I could see two crews of goggled men in khaki work clothes with various kinds of landscaping equipment. They were zipping weeds, buzzing back brush, and hauling off scorched limbs. I pulled over to watch and walk around for a while.

The more I learn about the unspooling of the ecosystems on our public lands, the more I'm convinced that at least part of their future involves more gardening and zookeeping. I don't mean everywhere, but certainly in places where we might have big positive impacts. And those must include what are now some of our most protected places: wilderness areas and wildlife refuges.

This should be done with sobriety and care. The planning framework outlined in chapter 3 gives us a template. A national inventory of vulnerabilities and opportunities comes first. That part was being done but had been held back in the Trump administration. Setting priorities and selecting projects is much harder. That stage should be guided by the science but informed also by the needs of historically disadvantaged groups, particularly those of the Native tribes, whose lands are being vandalized by climate change.[46] Louisiana's approach to coastal restoration, discussed in chapter 6, shows the importance of continual monitoring and adaptive learning.

Do we need to change the law? I've talked to legal scholars about this, including those mentioned above, and I think the answer is yes. It is true, as Eric Biber and Elisabeth Long Esposito suggest, that with enough flexing and stretching our current preservation laws could accommodate the more intrusive interventions we need. But that's like saying that with enough training you can get an ostrich to do what a horse can. Maybe. But at some point, design and temperament win out.

Rob Glicksman pushed this point in a conversation we had following a recent podcast interview I conducted with him. He urged me to compare the Wilderness Act with the 1970 Clean Air Act, a statute that the U.S. Supreme Court said in 2007 authorizes the EPA to regulate carbon emissions. Climate change, of course, is not mentioned in the Clean Air Act, and few people at the time would have thought of something as plentiful as carbon dioxide as being an "air pollutant." Still, in that foundational case (known as *Massachusetts v. EPA*) the justices found needed flexibility.[47]

"The issue there," Glicksman told me, "was, Can EPA do something that Congress didn't anticipate because it wasn't aware of the problem? Is the statute broad enough to encompass regulatory authority over a problem that they never thought about? And the answer the Court reached was, well, yes, the statute defines pollutant broadly enough to allow them to do that."

But that's different, he said, from the Wilderness Act, which is not about prompting interventions of a certain kind but about affirmatively disabling them. "You've got an additional hurdle here," said Glicksman, referring to the Wilderness Act. "The statute's not only silent, it's *impairing*. It creates mandates that are to some extent inconsistent with what we need to do to adapt to climate change. And so it's a more serious problem." It's true that in the Obama administration, officials like me tried to push agencies like the Park Service to test the limits of noninterventionist mandates. On the other hand, as Glicksman also pointed out to me, President Trump's reversal on this score leaves us with no way of knowing whether the strategy would have worked and, crucially, whether the federal courts would have gone along.

Just as important as changing the law is increasing the resources within land management agencies and encouraging them to put a higher priority on building resilience. Remember those complaints from land managers in the Mountain West about resource constraints and the lack of public concern? That's where citizen commitment is critically important.

Ideally, agency staff at individual parks and refuges would feel emboldened to press their superiors for whatever scarce resources might be available for growing adaptive strength. The White House would direct agencies by executive order to ambitiously pursue adaptive strategies as far as their statutory authorities allow. At the same time, Congress would work to add adaptive directives into the major land-management statutes and give agencies the flexibility to pursue a science-driven course, all with opportunities to reevaluate and change direction if needed. The White House would support budget increases to support this work, and Congress would assent. Any of these developments would be better than what we have. But none of them will happen without citizens at the local, state, and national levels pushing for change.

A great way to start, I think, is for people to find a piece of federal land they care about—a park, a forest, a wildlife refuge—and learn about how climate change is affecting it. They should visit the land and specifically ask staff members there about climate change and the options for building resilience. The rangers and other officials I talk to say they are eager to talk about the climate crisis, but they are concerned visitors will be put off by the bad news. They end up thinking that people don't really care. That's tragic, because, as we've seen, land managers take public opinion very seriously and use it to inform their priorities. People should share what they learn with others who might feel the same way—family members, fellow hikers, hunters, and fishers. Small actions like these can lead to more organized approaches like joining a nonprofit conservation group devoted to a particular landscape (the Joshua Tree National Park Association is a good choice) or learning about and commenting on new management plans when they are proposed.

When the *Los Angeles Times* tells us to "see those Joshua trees," we really should. Not because we're afraid we won't see them again, but because we want to make sure that we *will*. So it seemed to me that afternoon on the Lower Covington Flats, as I stood in the mud watching a crewmember drag a heap of branches to the road. He looked up, and through his safety goggles we made eye contact. I smiled and waved. "Take it easy," I shouted. He waved back.

It was an odd thing to say, I know. But that Eagles song still hung in my head and the whole scene was so surreal and sad—and yet not, I should emphasize, completely without hope—that it was hard for me to stay present in the moment. "Don't let the sound of your own wheels," I sang to myself, "make you crazy."

I kicked the stump of a withered yucca and I tried to imagine the causative conflagration—the bright embers whipping up, rabbits bolting for cover, an unimaginable, furnace-like heat coursing through the valley. The day before, I had asked Jane Rodgers why we shouldn't just give in to the landscape's "new normal." What would it matter, I asked, if there were no more Joshua trees in Joshua Tree National Park? Rodgers took on an officious air, reciting by heart

Congress's charge to "retain and enhance opportunities for scientific research in undisturbed ecosystems." I waited, and then her face softened. Besides all the needs of science, she said, it seemed important that people be able to experience this unusual and captivating plant in the place it calls home. That was, after all, Minerva Hoyt's promise.

"They could have called it Big Boulder National Park," Rodgers said. "Then it wouldn't have mattered so much."

chapter 10

The Octopus's Garden

We've talked, so far, about wetlands and mountains and deserts and more. But for a book titled for an octopus, I must also steer toward the ocean. I was sure Kathleen Sealey, the marine biologist, I wrote about in chapters 1 and 3, could help. So I reached out again to her by video from Whidbey Island, Washington. It was the same week I had been paddling in Useless Bay during the Labor Day fires, which I discussed in chapter 8.

I asked about the general health of octopuses in southeastern Florida.

"Well, I don't think anybody really knows. There's no formal assessment. Probably a lot less habitat for them than there was in the past."

"And what about warmer ocean temperature?" I asked. "How does that affect octopuses?"

It's a problem, she said. Like people, octopuses function best when their bodies are at a certain temperature. Unlike people, their bodies have no thermostat. As with fish and reptiles, their temperature fluctuates with the surroundings. Warmer temperatures limit how much energy the octopus can put into growing. "That's why tropical octopus are smaller than cold, temperate octopus," Sealey explained.

The ones in the Puget Sound, I suggested, must be a good size. There, water temperature hovers around 50 degrees F. Sealey told me that the biggest octopus she had ever seen was off the coast of Port Townsend, Washington. She found it suckered onto a Chevy Impala that had recently tumbled off a ferry dock. "The car was way down there—about forty feet," she said. "And there was that octopus just spread from the end of the trunk to the hood." The point being that an octopus's fate is, in the end, tied to temperature. While they do fine in tropical climes—conserving their calories and limiting their mass—there is a limit to what an octopus can tolerate. When the water gets too warm, they will eventually leave.

Warmer water temperatures are also threatening corals whose reefs provide octopuses and countless other creatures a rich habitat. A distant Beatles lyric bubbled into my head. "That's the 'octopus's garden,' isn't it?" I said to Sealey, "'The *coral* that lies beneath the waves.'"

"Yes," Sealey laughed. "Octopus want *structure*. They want *crevices*."

"A little hideaway beneath the waves."

"Exactly."

Unfortunately, coral—the Eden of octopus and reef fish alike, are in sharp decline in the United States and around the world. Many scientists now believe that unless we take immediate action, the very existence of coral reefs may be in jeopardy. Many of us know something about coral from the city aquarium or holiday snorkeling trips. We know that coral is remarkably beautiful. We may have a vague sense of its ecological importance. But many people couldn't say whether coral is animal, vegetable, or mineral. (Don't worry, Aristotle had trouble with this too.)

Sylvia Earle, the celebrated marine biologist, points out that even scientists know much less about the oceans than any other habitat. "Despite the swift advances in knowledge about the earth in recent years," she wrote in her 1995 memoir, "the fact remains that below the depth where most divers can venture, about 150 feet, little of the ocean has been seen, let alone explored."[1] We seem more captivated with outer space. While tech billionaires scout for new extra-planetary frontiers, we should be restoring Life's *oldest* frontier—the one stretching thousands of leagues under the seas.[2] (I've thrown shade on the tech barons, but as we'll see, there's one who deserves our praise.)

The ocean's role in managing climate and sustaining human life is unparalleled. It stores heat, shapes the weather, and loads oxygen into the air. Mapping the undersea world of climate resilience would require a separate alcove in Captain Nemo's brocaded library.[3] That is more than I can do in this book. But by casting a light on coral reefs and their vulnerability to climate change, I hope to show how topside and underwater ecosystems are connected and how we might build resilience in both.

Sealey says the coral crisis surprised even her. "I mean, I think if I had known that coral reefs would change so much in my lifetime, I would have paid more attention." Sealey's first job was as a researcher on Lizard Island, located on Australia's Great Barrier Reef. Back then, she said, "there's no way you could have told me that that coral would disappear in forty years. There was just so much coral. I was working on a project about fish and starfish then. To me, the coral was just kind of in the way."

What happened? Carbon dioxide emissions, which heat the oceans and alter its acidity, are part of the story. But so is offshore pollution, which comes in the form of agricultural runoff and dumped sewage. Then there is overfishing, which decimates plant-eating reef fish that protect corals from invasive seaweeds. In this way, the coral crisis mirrors Louisiana's wetlands crisis, which is also a story about poor land-based water management, overuse of a natural resource, and greenhouse gas. In Louisiana, the main culprits were restrictive levees (water management), unchecked fossil-fuel extraction (overuse), and sea level rise plus storms (greenhouse gas). In reef systems, the culprits are agricultural runoff plus human sewage (water management), unchecked fishing (overuse), and heat plus acidity (greenhouse gas). The moonshot prescription will be similar. Cap greenhouse-gas emissions pronto and buy time with a resilience program that reforms water management, checks overuse of resources, and restores what we can. Some of this can be done at the local and national levels. But it will require global cooperation too.

"All right, let's get serious," I said to Sealey. "Imagine it's 2050 or 2060. Give me your best guess: Do we still have tropical coral?"

"Umm. Maybe in aquariums?"

Ten years ago, I took up a new sport: recreational scuba diving. For as long as I can remember, I had wanted to try it. As a kid growing up in the 1970s in a Las Vegas subdivision—where tumbleweeds really *did* sometimes wander across the road—this might seem far-fetched. But watching all those episodes of *The Undersea World of Jacques Cousteau* on my grandparents' black-and-white television set really did a number on me. I wanted water. All around me.

It took a while, but finally on a vacation trip to Belize, I got trained and made my first dives on the Mesoamerican Reef, the second largest coral reef system in the world (Australia's Great Barrier Reef is the biggest Kahuna). While I don't dive very often, I do love it. But I feel a little late to the party. How long will the beauty of the ocean hold out? I decided that if I was going to take up scuba diving, I would have to see the activity as more than sport. The coral and reef fish were done lounging about, waiting to pose for holiday divers like me. They were fighting for their lives. And it was time for me to join the cause. First, I needed to learn more about corals—their history, their internal workings, their surroundings. I was surprised to realize that although I had been teaching for twenty-five years about environmental law and natural resource policy, much of the time on the Gulf of Mexico, I knew surprisingly little about ocean ecosystems. I could rattle off names of cacti in the Mojave, and I knew my way around a cypress swamp. I had hiked among goats and marmots in the North Cascades. But I had yet to submerge myself into the remaining 70 percent of Earth's habitat, the part on which *all* of life depends. I was one of those terrestrialists that Sylvia Earle complains about.

So, I contacted Roxane Boonstra at the Coral Restoration Foundation, a nonprofit marine conservation organization headquartered in Key Largo, Florida.[4] CRF, the world's largest organization of its kind, is dedicated to restoring coral reefs to a healthy state, in Florida as well as globally. It offers public educational programs, engages innovative coral research, and operates one of the most highly regarded coral restoration initiatives in the world, about which we'll learn more later.

Boonstra, who manages CRF's dive training, understood my story. Few people, she told me, take up diving for the activism—it's the beauty and mystery that makes them bite. But once the hook is set, many become stewards of the sea forever. Boonstra invited me to Key Largo to attend a typical training course for recreational divers who want to take part in the coral recovery program. After that, we will don masks and fins and explore one of CRF's offshore coral nurseries, where the corals are grown before "out-planting" them into recovering reef systems. "I'll book a flight down," I said.

And that's how I took the plunge.

Our day starts in a modest classroom building situated near the eastern shore of the island of Key Largo. There are instructive posters pinned to the walls, along with a couple of dog-eared maps. In front is a whiteboard and flatscreen monitor for showing PowerPoint slides. In the back, there is a large gurgling aquarium filled with several kinds of bright-hued corals. A few folding banquet tables line the perimeter, on which rest various samples of fingerlike reef skeletons. There's a rubber hammer and a pair of cutting pliers too.

Boonstra is our instructor today. She will be teaching us about coral in the morning and guiding our underwater tour in the afternoon. Boonstra is an accomplished diver with a master's degree in coral biology. A Miami native, she says some of her earliest memories are of snorkeling at John Pennekamp Coral Reef State Park in Key Largo. At ten, she learned to scuba dive there. For the better part of thirty years, Boonstra has dived everywhere in the Florida Keys and on many reefs around the world. And in those thirty years, Boonstra says she has watched the majority of coral in her home state "disappear in front of my eyes."

There are about fifteen students in the room, mostly white, evenly split by gender. The youngest is a boy—about ten years old—who is sitting behind me with his mother. There is a group of college students to my left, their notebooks open and gel pens poised. The man sitting next me is a retiree from Chicago who volunteers as a diving docent at his city's renowned Shedd Aquarium.

Boonstra opens her lecture with the classic ice-breaker question: Are corals animal, vegetable, or mineral? Even in this group of divers, not everyone knows. And, as I said before, even Aristotle—who is generally considered to be the West's first known marine biologist—was baffled, eventually concluding it was a "plantlike" rock.[5] Despite all that, Boonstra assures us that corals are indeed *animals*, though some flirt at the line between plant or rock.

Formally speaking, corals belong to the phylum Cnidaria ("nih-DARE-ee-uh") which also includes sponges and jellyfish. The phylum's calling card is a network of specialized cells used for catching prey. These cells, which usually contain neurotoxins, mean two things: First, they can sting you! Second, all those unusual chemicals make for a pharmaceutical cornucopia, whose bounty is disappearing faster than we can catalog it. For this reason, Boonstra says, corals are known as the "rainforest of the ocean."

Corals live in colonies. These are the waving fans or stony antlers that most people have in mind when they think about coral. In contrast, the animal, itself, is a tiny polyp anchored in a rigid cup. The polyp has its own mouth, its own stomach, and its own fluttering tentacles—juiced up with disabling chemicals—to seize and immobilize prey. The polyps are connected to one another by a very thin layer of living tissue. If one polyp catches food, it can share the nutrients with other neighbors in the colony via this tissue, like a "backyard barbecue," Boonstra tells the class. Together, these polyps form a kind of superorganism.

We call corals either soft or hard. The soft corals live in fleshy colonies that look like plants or trees. The secrets of these fauna we must leave for another day, Boonstra tells us, for we are concerned only with reefs. Reefs are built by what are known as hard, or "stony," corals. These corals—with names like "pillar," "boulder," and "staghorn"—are often named for the shapes of their rigid skeletons. Each polyp contributes to a skeleton by secreting calcium carbonate, a versatile mineral used by nature to build all sorts of things, from limestone to snail shells to human bones. When the skeletons fuse or overlap with the skeletons of other colonies, they form magnificent reefs, some on a colossal scale. Australia's Great Barrier Reef, which is really a system of some three thousand coral reefs and nine hundred coral islands, extends for more than 1,200 miles and can be seen from space with the naked eye.[6] The Mesoamerican Reef stretches more than six hundred miles through the waters of Mexico, Belize, Guatemala, and Honduras.[7]

Boonstra says soft corals first appeared in the Cambrian Period about 535 million years ago and hard corals in the Triassic Period about 240 million years ago. (We *Homo sapiens*, by contrast, have been around for less than 300,000 years.) Boonstra asks us to guess the age of the oldest living coral in the world. The boy sitting behind me says a thousand years, which I think was a good guess. After all, I know of bald cypress trees in the southeastern United States that are pushing 2,700, and I've seen bristlecone pines in my home state of Nevada that are thought to be more than 3,500 years old. But a colony of deep-water coral in the Hawaiian Islands has those terrestrial Methuselahs beat. According to Boonstra, recent radiocarbon dating pegs the age of that soft, feathery specimen at a whopping 4,265 years. "That coral," Boonstra says, "was born right around the time that the Egyptians kicked into 'Great Pyramid' mode."

I think about the Joshua trees I wrote about in the previous chapter. Living a long time in a challenging environment requires lots of energy, which often translates into very slow growth. I ask Boonstra if anyone knows how fast these corals grow. Using sophisticated carbon-dating techniques, scientists have, in fact, measured that, she says. The ancient corals of Hawai'i are believed to grow no more than thirty-five micrometers a year—about the width of a human hair.

To grow their skeletons, hard corals extract calcium carbonate from minerals dissolved in sea water. But our greenhouse-gas emissions are reducing the supply. The oceans, Boonstra explains, absorb the bulk of atmospheric carbon dioxide. The added carbon dioxide increases the acidity of the water, which is another way of saying that it depletes (by way of a chemical reaction) the amount of calcium carbonate available for extraction. When the supply of calcium carbonate falls, corals have trouble building sturdy skeletons. For similar reasons, she says, marine scientists are now finding sea snails with thinner and more fragile shells.[8]

In addition to the relationship between animal and mineral, there is another bond we must understand to appreciate coral's climate crisis: the animal's intimate dependence on vegetable life. Having marveled earlier at the bouquets of venomous coral tentacles in the classroom aquarium, I was disappointed to hear that, despite all that weaponry, most hard corals "are pretty terrible at catching food." It turns out that being stuck permanently at the same address comes with a cost. Therefore, Boonstra tells us, corals have devised a workaround, called "zooxanthellae symbiosis." Basically, a coral polyp invites a species of algae, called zooxanthellae ("ZOH-uh-zan-THEL-ay"), to live in its tissue. There the tiny plant cells enjoy protection from predators and reliable access to sunlight, which (through the process of photosynthesis) they use to make food (glucose, glycerol, and amino acids). The algae take some of these nutrients, but the bulk is passed on to the hungry polyps which grow, multiply, and build more reef. As a bonus, the algae, whose vibrant colors shine through the polyps' translucent tissues, bestow upon healthy reefs an unmatchable cinematic brilliance.

If all of this sounds a little strange, Boonstra assures us that such landlord-tenant relationships are all around us. To name one example, scientists tell us you have about five hundred different kinds of "gut bacteria" living in your intestines right now, along with vast communities of viruses, fungi, and who knows what else, all of which help to break down the food you eat into the chemicals your body needs. That may not deliver cinematic brilliance. But suffice it to say that without all those little critters, you would starve.[9]

Now, here's the connection climate resilience: It turns out that while coral and zooxanthellae get along well, the relationship is taxed when corals get stressed out, particularly when there is a spike in water temperature. For reasons scientists are still trying to understand, this can cause the polyps to expel the zooxanthellae. The corals turn a ghostly white, and they begin to starve. That's what "coral bleaching" is. Scientists say it takes fifteen to twenty years for corals to recover from bleaching and become healthy enough to support marine life again. But with all the pressure bearing down on coral reefs, it's harder for them to rebound.

"So, you've got a bleached coral because of higher temperatures," Boonstra says. "And then it can't build its skeleton because of ocean acidification. Now you have *two* stressors." "Then you have sea level rise. All the corals need sunlight for the zooxanthellae to make food. But when the depth of the water increases, photosynthesis gets harder." That's a *third* stressor, she says. "And then you've got more intense and destructive storms." Our tiny cnidarian polyp is now subjected to *four* stressors related to carbon-dioxide emissions. And scientists are still counting.

Coral reefs occur in more than a hundred countries around the world. While they cover only 0.2 percent of the ocean floor (mainly in shallow, tropical waters), they provide outsized benefits to the human population. Their fish (a quarter of all marine species) supply essential protein to hundreds of millions of people, many of whom are of limited means living in poorer countries. Their limestone boulders and thickets protect coasts from storms. Their mystery and beauty support billions of dollars in tourism. This bounty totals $2.7 trillion per year in goods and services, according to a comprehensive report by the International Coral Reef Initiative, a partnership of countries and organizations that works to protect coral reefs.[10]

For all that, the world's corals, which are thought to contain as many as nine hundred species, are facing a nearly unfathomable existential threat, the product of many layers of stress, all caused by human activity. These include local pressures like coastal development, agricultural runoff, sea-based pollution, and overfishing, along with impacts related to greenhouse gases, such as marine heat waves, ocean acidification and more.

The coral crisis happened in the same way as Mike Campbell's bankruptcy in Ernest Hemingway's *The Sun Also Rises*: "Gradually and then suddenly." Near the end of the twentieth century, despite land-based pollution and aggressive fishing, the world's reefs were holding their own. Hard coral cover in some regions was declining, but the global average was a respectable 30 percent or more. Then in 1998, a marine heat wave triggered a mass coral bleaching event that killed 8 percent of the world's coral. (For context, that's more than all the coral now living on reefs in the Caribbean Sea.) In an amazing example of species resilience, reef corals around the world slowly revived and, within a decade, had jumped back to pre-1998 levels.[11]

That pattern did not hold. Since 2009, the world has seen more marine heat waves and more mass bleaching. These events, on top of local pollution and over-fishing, have brought coral to the breaking point. In the last dozen years, we've lost 14 percent of the world's reef corals. This time, because of the frequency of new heat waves, reef corals did not rebound. That's bad news. The sliver of good news is that in some places, such as East Asia's Coral Triangle Region, where 30 percent of reef coral resides, the reefs seem much more resilient and are even regaining lost mass. Scientists still don't know exactly why. The effect could be attributable to the mix of species, the genetic makeup of individual colonies, the local environment, or some combination. But many coral scientists believe that through a combination of emissions reductions, ocean protection, and coral res-toration, we have a fighting chance to save and restore a meaningful chunk of this underwater treasure. As scientists for the International Coral Reef Initiative conclude in their report, "Many of the world's coral reefs remain resilient and can recover if conditions permit."[12]

If conditions don't permit—that is, if we fail to limit global temperature increases to 1.5 degrees C and give up on protecting and restoring what we have—we could lose it all. Reef corals, which provide the foundation for one of the most complex ecosystems on the planet, would go extinct. Saving coral, it turns out, is a little like restoring Louisiana's fragile wetlands. We may never recover everything that's been lost, but we can protect and rebuild in strategic places designed to enhance selected benefits. We can "spot weld" and hope that some of the meshing holds.

In 2018, a group of twenty-one conservation scientists from around the world proposed just that. In a paper published in *Conservation Letters,* they outlined a strategy to identify and protect coral reefs "in the context of rapid climate change."[13] Knowing that no strategy could protect them all, they surveyed the world's oceans looking for regions where long-term investments in conservation would go the furthest. Specifically, they focused on areas where heat waves and cyclones were less likely to occur, where fatal diseases were less likely to spread, and where coral populations seemed most likely to rebound. The result was a list of fifty reef systems (each containing about two hundred square miles) that should be given top priority. Their portfolio included reefs in French Polynesia, the Caribbean Sea, the Red Sea, East Africa, the waters of Southeast Asia, and the whole of Australia's Great Barrier Reef.

The proposal, which defines clear priorities and incorporates state-of-the-art climate projections, surely wins points for being forward-looking. The authors also build in flexibility by inviting others to add their own priorities and pull new data into the mix. They also explain how this portfolio, designed for a global initiative could be fit into a smaller regional or national framework. What the authors don't do is take adequate account of the severe distributional effects such triage would have for populations whose reefs don't make the list. That would include millions in Central America who rely on the Mesoamerican Reef, an undeniably productive resource that, nonetheless, resides on a "hurricane highway." As with Louisiana's coastal restoration plan discussed in chapter 6, the aims of maximized utility can easily conflict with distributional fairness. Still, human benefits and cultural values could be factored into a program like the one these scientists propose.

Experts in coral policy generally agree on the basic strategies needed to save coral. The first, as the previous example suggests, is to learn more about what is actually out there. Having better information helps decision makers decide what is necessary and realistic to protect. It helps scientists learn more about what makes a particular coral species or genetic variety more adaptable. As recently as 2020, a team of French scientists discovered a sprawling coral reef resembling a bed of roses off the coast of Tahiti that is in pristine condition. Nearly one hundred feet below the surface, the reef, which extends 1.86 miles, was apparently unknown to anyone.[14] In a region known for declining corals, why has this flowerlike garden thrived? Are there other reefs like it? Can they be replicated? We can no longer afford to be kept in the dark.

Fortunately, several large efforts to inventory and monitor coral are shining light into the shadows. One of the most exciting was initiated by the late Paul Allen, the cofounder of Microsoft and philanthropist, who died in 2018. Developed by an international team of researchers and institutions, including the University of Queensland in Australia and the Carnegie Institute for Science, the Allen Coral Atlas offers the first global map of shallow coral reefs.[15] Managed by scientists at Arizona State University, the online platform continually updates regional coral maps to show events like bleaching in almost real time and to monitor other changes in surrounding ecosystems. The atlas, developed from more than two million satellite images, contains maps of nearly 100,000 square miles of reef, including outlines of marine protected areas, seagrass, rocks, sand, and other features.

Online and free to the public, Allen's atlas is deployed by countries around the world. The Solomon Islands and Fiji Islands are using the tool for developing marine protected areas, while the Bahamas are using it to select the best areas for coral reef restoration, where they "outplant" coral that has been grown in nurseries. Paul Allen never gave up his passion for sci-fi, space opera, and the lure of the stars. But in later years, he reached for the starfish too. Excepting Microsoft, the Allen Coral Atlas may prove to be Allen's most sweeping contribution to the world.

In addition, we need to clamp down on the non-climate-related stressors that are already making the lives of coral miserable. Relieving those stressors allows reef systems to devote more of their energy to build resistance to and or recover from climate-related stressors like heat waves and increased acidity. Relieving such stress buys time for us to reduce carbon emissions and for the temperature of the planet to gradually find equilibrium. The good news is that the biggest non-climate-related stressors—shoreline pollution and overfishing—are problems we already understand well. The bad news is that we haven't shown much resolve in fighting them. That needs to change.

The shoreline pollution of greatest concern is agricultural runoff and human sewage. Runoff from agriculture pushes tons of fertilizers and other "nutrients" into bays and other ocean systems where they trigger expansive blooms of algae and other plant life. Algal blooms can block a reef's access to sunlight, which, as we will learn, is important to food production. When the algae die, the process of decomposition consumes oxygen from the water, which the reef's animal life requires to survive. Laws regulating agricultural runoff are seldom as strong as they should be, including in the United States. Farm lobbies, the world over, are known for their political influence. In addition, the downstream harm is invisible to most of members of the public.

Human sewage pollutes coral reefs with disease-carrying pathogens as well as nutrients that cause algal blooms. It comes from shoreline septic systems, whose fluids leach into the water table, or from municipal sewer systems that intentionally or unintentionally expose ocean systems to waste. As Sealey told me during our conversation, "everybody's treated the ocean as this big toilet. There are too many people who put too much stuff into the ocean." Many people assume such problems are found only in poorer countries, but rich countries also send sewage into the sea, even though most could easily avoid it.

When can the ocean legally be used as a toilet? In 2020, the U.S. Supreme Court dived into that controversy in a case involving the pollution of beaches and coral off the coast of Maui, Hawai'i.[16] The county of Maui operated a sewage treatment plant about a half a mile from Kahekili Beach. Instead of fully treating the sewage the way many facilities do, the plant only partially filtered the waste through grates and rocks before injecting it deep into the ground. By design, the fluid would then travel a few thousand feet laterally through the groundwater and be released into the ocean. In this way, the county discharged each day four million gallons of poop into the shallow waters of neighboring communities. An organization called the Hawai'i Wildlife Foundation sued.

There were lots of problems related to these discharges, but for our purposes, let's focus on the fact that this waste—combined with that from other sources—has been degrading the coral of Hawai'i for years. Following an ocean heat wave in 2015, it is estimated that a quarter of all the state's coral reefs died from bleaching because they were not resilient enough to recover.[17]

In the United States, the discharge of sewage into the ocean must generally comport with federal pollution standards provided for in the Clean Water Act. This is the same statute, you may recall from chapter 6, that establishes standards for the dredging of wetlands in coastal Louisiana. Maui County, backed by the Trump administration's Department of Justice, maintained that those standards did not apply to its facility. The Clean Water Act requires a federal permit to add a pollutant to a navigable surface water from a "point source," which in this case would be the injection well.[18] In response, the county argued the sewage being discharged into the ocean was, in fact, not coming from a "point source." It was coming from the groundwater instead.

You don't have to be a lawyer to see the silliness in this argument. If I drop a flowerpot on your head from my balcony, it is surely being discharged from my hands rather than from the air above you. Justice Sonia Sotomayor smelled a rat. "What current regulations exist that stop the county from polluting the ocean?" she asked in oral argument. "It's definitely happening, and Maui County says the Clean Water Act shouldn't stop them—so are they just going to get away with it?[19] In the end, a 6-3 majority ruled against the county, finding that the discharge amounted to the "functional equivalent" of a direct discharge from a "point source."[20]

I mention this example for a few reasons. First, it illustrates the difficulty—in the United States and elsewhere—of regulating fluid discharges once they leave a pipe or other "point source." That is one reason that agricultural runoff and sewage are so often able to escape even laws that are on the books. You might think that the source of sewage flowing along Kahekili Beach should have been obvious, but in reality, the source and volume were only verified after professional analysis that included tracing the flow of dyes in the water. Aside from the political pressures exerted on behalf of farmers and residential developers, enforcing rules against this kind of pollution can be expensive and labor-intensive.

Second, I can't overemphasize how trifling some very important questions might seem before you see their connection to the wider world. This happens all the time where climate resilience is concerned, since few people in the past set out to write rules or policies specifically about that. If before reading this chapter, someone had told you the crux of Hawai'i's coral resilience policy hinged on the meaning of the word *from,* you would have rolled your eyes and ordered another Mai Tai. But this stuff matters, in part because climate resilience disputes are new and our legal and policy language has yet to adapt.

Finally, as the previous point suggests, if our statutes and regulations are to prove flexible enough to protect us from climate breakdown, we need judges who see the law in *functional* terms. You might think that goes without saying, but in the *Maui* case, three justices treated the issue like a formal grammar exercise and ruled the other way (although I think even the grammar in this case favored the environmentalists). One of the "functionalists" in the majority, Justice Ruth Bader Ginsburg, has since died and been replaced with Justice Amy Coney Barrett, who is generally viewed as a "formalist." Formalists ask only, "What does *from* mean?" Functionalists, in addition, ask, "Are they just going to get away with it?" We need more of the latter.

In the previous chapter on public lands, we saw that one potentially powerful tool for building ecosystem resilience involved carefully developed limits on use. In a country like the United States, where the federal government owns 28 percent of the land, limits like that can have a big impact. This lesson applies to coral ecosystems too. Generally speaking, international law recognizes that maritime nations have territorial control of their waters from their shore out to a distance of twelve miles. Within that zone, countries can harvest fish, extract resources, and (with some exceptions) control sea passage as if the zone were just another piece of land. Maritime countries also have claim to something called an "exclusive economic zone," which is made up of the water column and the seabed out to a distance of two hundred miles. (Overlapping areas are often divided equally among the parties.) In that zone, a country cannot control sea passage, but it can control fishing and other resource extraction.[21]

While it's true that most of the ocean lies outside either of these zones, the good news is that coral reefs do not. Because most hard corals prefer shallow waters close to the sunlight, nearly all the world's important reef systems lay within the regulatory control of national governments. (Their proximity to shore also puts them at risk to the land-based pollution discussed above, so this sword definitely has two edges.)

This has led to an international trend toward so-called, "marine protected areas," or MPAs. These are like the national parks or conservation areas we have on land, a kind of "zoning" for the sea. They have boundaries that are clearly defined, are managed under formal laws or cultural norms, and seek the long-term conservation of natural systems.

Like the land-management units we saw in the preceding chapter (national parks, national forests, BLM land, and so on), MPAs come in different flavors. Some have "no-take zones," barring all fishing, drilling, and other extractive activities. Some limit these activities only for certain actors or under certain conditions. Some MPAs are managed at the local level with strong community involvement. Others are not.

The United States maintains a system of National Marine Sanctuaries and Marine National Monuments that encompass more than 1.5 million square miles of territorial waters, an expanse larger than India. These waters extend from Washington State to the Florida Keys, from Lake Huron to American Samoa. Most of these areas were designated by the National Oceanic and Atmospheric Administration (NOAA) according to a 1972 statute passed after the 1969 Santa Barbara Oil Spill. A few parks, like the mind-bendingly extravagant Papahānaumokuākea National Marine Monument in the Hawaiian Islands, were created by presidential decree under the Antiquities Act of 1906—the same law that allowed President Franklin Roosevelt to create Joshua Tree National Monument. (You can thank President George W. Bush for that Hawaiian monument and President Barack Obama for expanding it.)

There is strong evidence that MPAs, when well-crafted, strengthen climate resilience on coral reefs. Several side-by-side comparisons of heat-wave effects on protected and unprotected waters in the Pacific, for instance, have shown total recoveries on fully protected reefs, in contrast to mass mortality on unprotected reefs.[22] As Sylvia Earle, a staunch advocate for MPAs, explains, "There's plenty of evidence – plenty of evidence – that when you stop killing things, recovery can happen."[23]

The trouble is that as of 2017, only 3.5 percent of the ocean is covered by MPAs; and only 1.6 percent is covered by MPAs with no-take provisions.[24] Despite the recent increase in MPAs around the world, we are a long way away from the United Nations' target of having 30 percent of the ocean protected by 2030. Add to that the fact that many MPAs—including those in the United States—fall short in terms of protective standards and adequate enforcement. Researchers find that the biggest barrier to success with MPAs worldwide is the lack of adequate staffing and funding.[25]

Critics say many MPAs amount only to "paper parks"—areas protected on paper, but not in real life. They suggest MPAs may not be worth all the time and expense to create. Earle says that's short-term thinking. "Marine protected areas do, in fact, work," she said in an interview with the media project Oceans Deeply. "Teddy Roosevelt faced a similar problem when he began, in the early twentieth century, establishing protections for national parks and monuments in the American West. There was a lot of poaching, a lot of tree cutting. It took a while for people to realize that they are really harming their own interests when they take from these safe havens for nature." Putting a park on paper is, of course, not sufficient, but Earle says it is necessary. "You have to first establish it before you can protect it. If you don't establish an area and formalize the nature of a place as, call it an MPA or call it a reserve or call it whatever you want, you can't protect what doesn't exist. That's what Roosevelt found. That's what we're finding now."[26]

To review, fighting the coral crisis requires cutting carbon-dioxide emission, strengthening land-use controls to reduce agricultural runoff and the discharge of sewage, and clear limits on fishing that are adequately enforced. The road to success will depend on the country involved, but a proper foundation must always include reliable, updated information and financial resources. The enforcement staff and other infrastructure must be paid for. And some parties who stand to lose in the short term—subsistence and professional fishers, oil rig workers, and the like—must be compensated or retrained for other rewarding jobs. It's the practical and moral thing to do. To accommodate this, wealthy countries may have to rearrange some financial priorities. Poorer countries must as well, but they also deserve financial backing from the rest of the world, which, after all, contributed much more to the climate crisis, and which will benefit enormously from coral resilience in the tropics.

Sylvia Earle asks us to play the long game, and she's right. But she, better than anyone, knows that coral reefs don't have much time left. That's why, in addition to these top-down, policy-driven approaches, we should also support efforts that take a more bottom-up approach. One of the most promising may be coral reef restoration of the kind that CRF and other organizations around the world are now promoting.

Boonstra has asked us to stand up, stretch our legs, and assemble around the "coral tree," which stands in a corner near the aquarium. Made of fiberglass and white PVC pipe, it stands about six feet tall. Its "trunk" is a round tube three or four inches in diameter. Every foot or so, a series of perpendicular "branches" extends out. Someone says it looks a like a Christmas tree. Maybe, but more the weepy, "Charlie Brown" kind, I think.

Each branch has several chalky, finger-sized fragments of coral skeleton hanging from it, each suspended by a nylon fishing line. The skeletons we see are from staghorn corals (whose spindles resemble the offshoots of deer antlers) and elk-horn corals (whose flattened limbs remind me not so much of an elk but of Bullwinkle the Moose). "Coral trees" like this were developed by CRF a decade ago to allow for growing starter corals in offshore nurseries. They are now widely recognized as one of the best methods for growing large numbers of certain species of corals and used by groups around the world. Fragments of staghorn or elkhorn are snipped off more mature corals (originally taken from the wild, but now from the nurseries themselves) and then tied to such trees, which are suspended from the bottom of the ocean in orchard rows. (Each is crowned with an air-filled buoy to keep it standing tall.) The trees in the nursery are anchored in about thirty feet of water. Each tree holds about sixty corals.

Then Boonstra clicks on a slide on the flatscreen monitor and a vibrant stand of coral trees materializes. Under swaying waves, festooned in polychromatic living coral, the orchard is a showstopper. "This is our Tavernier nursery," Boonstra says, referring to the foundation's most well-known nursery located off Key Largo, the one we will dive in. "It has more than five hundred trees total. That's about thirty-five thousand corals growing in an acre and a half of space. There is nowhere else in the world that you can swim through a living forest of coral like you can in Tavernier. And that's just one." In all, CRF maintains eight nurseries, she says, all in Florida. Together, they house almost fifty thousand corals that can be reef-ready every year.

These are staghorn and elkhorn. They are pretty hardy, and they grow quickly, meaning that in less than a year a fingerling on the tree can be harvested and out-planted. Boonstra likens these corals to the underbrush of a terrestrial forest: they move in fast, stabilize the soil, and set a structural foundation. Something like the marsh grass and wax myrtle you see in Louisiana's Maurepas Swamp, I thought. But you also need the equivalent of massive cypress and tupelo trees to provide shelter and defend against storm and surge. So the nurseries also house more than a dozen other species, including globe-like boulder corals to flesh out the habitat.

In addition to species diversity, CRF is experimenting with genetic diversity too. Just as some people have a higher tolerance for heat or a stronger immunity to disease, individual corals within a species can vary significantly in their ability to survive external stressors. Scientists at CRF select to maximize genetic diversity and, in particular, to fill the area with individuals whose genetic package (called a "genotype") has proven to be especially resilient in the wild. Divers return at least fifty different genotypes of staghorn and fifty different genotypes of elkhorn to every restoration site. According to Alice Grainger, CRF's director of communications, there is now more genetic diversity in the organization's nursery programs than can be found in the wild.

The nursery requires regular maintenance. Divers work in shifts to brush invasive algae and other irritants from trees to remove any competition for young colonies. As the specimens grow, they must also be "fragmented." That's what the cutting pliers sitting on the classroom table are for. Divers will prune the branches of healthy corals to create more "starts" that can be grown on another branch.

When the corals are large enough, they are then out-planted onto degraded reefs. Workers collect specimens from the nursery, load them into tubs and milk crates, and truck them down Florida's Overseas Highway where they are used to repopulate reefs along the upper and lower keys. At the reef, divers delicately clear space and glue the corals in place with marine epoxy. On a single tank of air, a trained diver can plant anywhere between fifty and one hundred corals on a reef, depending on the species and the conditions in the water.

How does this work at scale? "With hundreds of people with thousands of corals," Boonstra beams. Every year, the foundation hosts an event called "Coralpalooza" in honor of World Oceans Day. In 2019, 250 people out-planted 1,760 corals in a six-hour period for that event. For that whole year, foundation volunteers planted 30,000 corals. Survivorship is high, about 77 percent.

<hr />

Floating above the Tavernier Nursery dive boat chartered through a local diver operator and staffed by the CRF team, my classmates and I prepare for our first dive of the afternoon. People are wriggling into their buoyancy jackets, checking airflow, spitting into their masks to prevent lens fog. I mentally tick off the rules Boonstra has taught us for visiting the nursery: don't touch the fragile bottom; move slowly around the coral trees; use "fin awareness"; don't get staghorn caught your hair; and don't use sunscreen containing oxybenzone or octinoxate, which kills baby corals!

Someone asks about gloves to avoid stings. Many marine areas forbid the wearing of gloves—the idea is that you're more likely to touch things you shouldn't be touching. But volunteers charged with brushing and fragmenting corals are *supposed* to be touching things. "Wear gloves for sure," Boonstra, says, "otherwise, the fire coral will do a number on you." I pull on my own pair of stretchy reef gloves, the kind with suede padding stitched onto the fingers and palms. One by one, we take backward somersaults from the side of the boat and fall into a magical world.

I really was not prepared for how peaceful and beautiful this artificial garden would be. Thousands of vibrant corals swayed around us from every height. That I expected. What I had not considered was the volume and variety of sea life that such a contrivance would attract. To my left, an undulating tree of orange and brown wrapped in a garland of silvery fish. To my lower right, a passel of hogfish—their red eyes afire—rooting for stone crab along the sandy bottom. Someone pointed to the outline of a large olive shell bobbing several yards away. I learned later that was "Hubie," the resident green turtle, who paid a visit to Tavernier months ago and never left. The biodiversity in this nursery is thought to equal or surpass the biodiversity of nearby natural reefs. One recent survey showed that more than ninety different species had taken up residence in the surrounding area.

I gently sweep my body into a passage between two rows of trees (fin aware-ness!). To get a better view of the corals, I angle my body into a vertical position, head down and fins up. I take a bobbing fragment of staghorn in my padded hand. Every staghorn on this tree is of the same genotype, that is, having the same collection of genetic information. The limestone limbs are thick and stubby. The neighboring coral tree is also filled with staghorn corals. But the limbs of these specimens are long and branchy. The same species, but such different appear-ances. And that's just what you can see, Boonstra had told us in class. Think of the range of diversity inside that you *can't* see—differences in heat tolerance, metabolic rhythms, resistance to blight. There is a cornucopia of genetic diversity in these enchanted waters. Suddenly, I hear a low, percussive sound from below. It's eerie and loud, like a kid eating Cap'n Crunch cereal at the bottom of a well. I look over and see a green-and-blue parrot fish scraping algae from a coral branch with its tough, beaklike mouth. Another nursery volunteer, I think. We need all we can get.

———◆———

Today, there are literally hundreds of volunteer-based coral restoration projects around the world—in Hawai'i, Mexico, Bonaire, the U.S. Virgin Islands, Gre-nada, Thailand, Malaysia, Indonesia, and many other places. But not everyone is as enchanted by coral restoration as are Boonstra and her army of volunteers. A familiar complaint is that coral restoration siphons funding from more top-down approaches and, despite CRF's impressive record, cannot restore reefs at the scale and in the time frame necessary. In addition, the quality and longevity of these programs vary enormously, as one would expect in such seagrass-roots effort. A global study of such efforts published in 2020 found that many pro-grams suffer from either a lack of properly aligned objectives or insufficient monitoring and reporting. [27] One might, of course, say the same thing about government efforts to control land-based pollution and operate marine parks.

When I asked Kathleen Sealey about restoration efforts, she was not encour-aging. "I mean, honestly, I don't think these coral nurseries are going to cut it," she said. "I feel like we're not dealing with the larger systemic issues, like water quality and land-use conversion."

But others, like marine biologist Maria Anderson at Ocean Group Maldives, are more optimistic. "Both large-scale efforts to address the climate crisis and labor-intensive replanting efforts are necessary to give reefs a chance of surviving

Earth's current extinction crisis," she said. "I've heard of hundreds of restoration projects around the world, but none that have failed."[28] Peter Harrison, a marine biologist at Southern Cross University in Australia, is pioneering the use of robots to distribute baby corals onto damaged reefs, doing a job in six hours that would take human volunteers at least a week.

New restoration techniques and the availability of enthusiastic volunteers have launched a variety of innovative policy ideas. One of the most fascinating is a coral reef "insurance policy," which was jointly developed by the Nature Conservancy and Swiss Re, a global insurance company. Remember the Mesoamerican Reef, the world's second largest coral reef and the one where I first learned to dive? It's a beautiful, highly productive reef that buffers the Yucatán Peninsula from hurricanes and storm surge. It also indirectly contributes billions of dollars each year to the tourist economy in the Mexican state of Quintana Roo, home to the resort cities Cancún and Tulum. You may also recall, that because of the increased risks associated with hurricanes going forward, at least some experts—like the authors of that report in *Conservation Letters*—doubt that a risk-prone reef like this deserves a high priority for saving.[29]

Now this limestone colossus is protected by the world's first and so far only insurance policy designed for a natural structure. Here's how it works. The state of Quintana Roo buys a policy to cover the loss of storm-based reef destruction. The payout is not based on the monetary value of reef damage. While theoretically possible, no one wanted to go through the hassle of verifying a number that complicated. (If you've ever made an insurance claim for wind damage on a house, you know what I mean.) Instead, the policy is triggered by something far easier to measure: the storm's wind speed. The stronger the measured wind speed, the worse the assumed damage to the reef. In this case, the reef is insured against hurricanes with a wind speed of a hundred knots or greater, the speed of a Category 3 storm.[30]

The policy's first test came in 2020 when Hurricane Delta slammed into the reef, triggering an immediate payout of roughly $850,000. After the storm dissipated, a team known as the Brigade sprang into action. The group, made up of diving instructors, park rangers, fishers, and researchers, descended on the reef. Armed with full diving cylinders and lots of epoxy, the team spent hours underwater fastening pieces of the reef back together and collecting fragments to seed new colonies.[31] Whether coral insurance policies and the restorations they fund will prove effective in the long term remains to be seen. But we will likely see more such instruments as communities further assess the monetary value of the imperiled ecosystems they rely on. A coral reef today, perhaps a desertscape or cypress swamp tomorrow. Our future will be full of mini moonshots.

If you've ever doubted what a small group of thoughtful committed citizens can do, you should spend more with people like Kama Cannon and Kara Norman. Cannon is the founder and director of education at an educational nonprofit called DiveN2Life, based in Summerland Key, Florida, about twenty miles east of Key West.[32] DiveN2Life offers an extraordinary program that instructs adolescents and young adults in scuba diving and scientific research. In the last seven years, DiveN2Life has introduced scores of young scientists to projects related to coral monitoring, reef seeding, and coral restoration. Her students, who range in age from eight to eighteen, perform real scientific research, working with universities, state agencies, and NOAA.

Kara Norman is one of Cannon's students. Norman is a fifteen-year-old high school student in Key West. Already a certified master diver, Norman is one of the youngest research divers in the United States. In fact, she is currently serving as the lead researcher, the principal investigator, for a peer-review study on the seeding of coral larvae on artificial reefs. Norman's passion for diving and coral restoration has taken her across the Keys and through much of the Caribbean.

When I first learned about DiveN2Life, I sought interviews with them both. One of my biggest interests in unpacking resilience policy is the effect it has on the people who get involved. You can have the best ideas in the world. But nothing gets done if people aren't energized.

Cannon says the secret to energizing students is to pay attention to the "identity" that is being formed. Cannon, whose research on scuba diving and place-based education for girls was published in 2019, told me that people assume many "layers" of identity in their lives and that these layers influence our interests and the endeavors we take on.[33] Cannon encourages students to nurture what she calls "a science identity." Helping adolescent girls to develop a science identity can be particularly important, Cannon explained, because the way science is taught in many schools does not support or acknowledge their interests. "Imagine a teenage girl. If she doesn't see herself as a scientist—a 'science girl'—she's not going to stick with the scientific disciplines. She's going to drop out of that pipeline and find something else."

I asked Cannon what draws her students to scuba and marine science. In the beginning, it's the magic of the ocean: the damsel fish, the sea turtles, the dolphins. For Norman, "the original draw was the diving." She wanted to get her certification. "I also saw pictures on Facebook about the science that they were doing. That interested me, but I wasn't super into it right away." Once she got more involved, "I realized how much I loved the science too."

Then, like most things worthwhile, the reward became more about connection and purpose. "We already know what makes kids engaged," Cannon told me. "It's the feeling that they belong, that their work is authentic, that it's meaningful—not just to themselves but to the greater community." "They love the fact that it's hard to do," she said. "It's real. We hand them clipboards and survey sheets and underwater cameras. They've got jobs to do." Here Cannon seems to be following the prescription advocated by climate scientist Katharine Hayhoe, whom we first encountered in chapter 5: "Bond, connect, and inspire."[34] The work might also involve travel to places like the U.S. Virgin Islands, an adventure for many.

Students' jobs might include collecting water samples or collecting data on coral growth, which might then be used by marine scientists at the Florida Keys National Marine Sanctuary or NOAA. Many of Cannon's students participate in coral restoration programs. Norman has spent a lot of time cleaning the corals in the nursery and "fragmenting" them underwater (figure 10.1). "We just go and get like one of the big corals chop it up into a bunch of little pieces, which sounds like super violent, but it ends up helping them grow in the long run. And then we restring them onto a whole new tree." She's done a lot of out-planting too, gently gluing coral fingers onto rock or popping baby boulder plugs into reef scalp. "It's definitely labor intensive."

How do Cannon's students react to seeing the rainforests of the sea in such decline? "They're definitely seeing change over time." Cannon told me. "They see the degradation of what they see as 'their' coral reef. They become very protective of it. So now the question is: Why does it matter to them?"

Cannon's answer: "Because they *love* it. They don't have the hopelessness of the political agendas or the economic issues. All they know is that they love it—and they are willing to fight for it. That's the most important part."

10.1 Research diver Kara Norman (*left*) and her colleague, Cameron Smith,
"fragment" branches of staghorn coral in a restoration nursery off the shore of
St. Thomas in the U.S. Virgin Islands.

(Photo by Kama Cannon; courtesy of DiveN2Life)

"There *are* days when I get discouraged," Norman added. "It's such a depress-
ing thing to think about. And I just feel so unmotivated because I'm doing
everything I can, and I still feel the whole rest of the world is falling apart on
these issues."

"I know what you mean," I said. "Sometimes it looks bleak."

"But then on the days when I'm doing hands-on stuff," she said brightly, "I see
that what I'm doing is really making an impact. Then, I think to myself, it may
not be a global-scale solution, but it's a little bit better than it was before."

I wondered, I said, how their volunteer work affects their political views.
Cannon told me that she does her best to surf above politics. She wants students
to know that "not everyone from a certain political party is the enemy." "In fact,"
she said, "most of the donors for my program happen to belong to a political
party that many young people say is ruining the world, but they're not."

"Look, I can't vote," said Norman. "I'm not on one political side or the other. What matters most to me is what leaders are doing for my future. I want them to have a reputation for protecting the environment."

This talk of the future made me wonder if there might be a tension between younger and older generations. "People say, 'Oh, wow, you're getting involved *so young*,'" Norman said. "I feel like I don't have a lot of choice." She speculates that people of my generation may have come to environmentalism later in life than young people today do.

Cannon said her students know that older generations have caused much of the environmental harm they experience. "But we also see the older group as mentors, because some of them are working very hard to address these problems," she said.

"I think it's super valuable that we are able to work together," said Norman.

I asked Cannon about the issue of scale. What does she say to the argument that only the big national or global initiatives are the ones that can move the needle? I could tell by her change in expression that I had touched a nerve.

"Look, the older folks who used to care about this stuff," she said, referring to some of the marine biologists who minimize out-planting efforts, "are starting to say all the money we're throwing at restoration is a big waste. Well, that is the *complete* wrong message to be sending to young, hopeful students." Cannon said the experience of out-planting and monitoring coral has been a "driving force" in propelling her students toward scientific discovery. "These are clever kids, the ones who have the intellectual capacity, the motivation, the creativity, the teamwork, and basically the passion to do something about climate change."

Before you know it, Cannon said, students get involved in public advocacy. They educate family members, their neighbors, and friends. Like the student advocate Wendy Gao, whom we met in chapter 7, Cannon's students have presented their research at schools, local hearings, and state meetings. The day I spoke with Norman, she had just submitted written public comments on a controversy in Key West involving the passage of massive cruise ships, which she says are "wrecking our ecosystem down here."

Years back, a colleague of Norman's made such an impression presenting on the Everglades that then-Congresswoman Debbie Mucarsel-Powell insisted the student diver accompany her to President Trump's 2020 State of the Union Address. Cannon's theory of layered identity, I began to see, didn't end with "science girl" (or boy). The layers continued to grow, including those of "protector," "educator," and, finally, "active participant in democracy."

Cannon put the issue in perspective: "Resiliency has to be woven into every subject in every school, because these kids are going to be the ones who put this world back together. I'm fifty-one years old," she told me. "I'm not going to be around when it collapses completely. What I *can* do is give these kids as many tools as possible in engineering and science."

I've become convinced that, in addition to top-down initiatives like pollution controls and MPAs, volunteer-driven coral restoration projects should be championed around the world. Strong research from the Florida Keys and many other regions shows that with the right techniques that restoration can expand coverage and buy time while coral reefs catch their breath. Combined with innovative insurance programs, like the one covering the Mesoamerican Reef, restoration could find a source of dedicated funding in certain circumstances. Equally important, I think, is the potential for volunteer-based restoration to fire the interest of young people around the world, to help them fall in love with the ocean that brought them life and commitment to protecting it. As we saw in chapter 5, few things get done in this world without sincere emotional attachment.

Still, I admit a small disappointment with my introduction to coral science and restoration. Throughout my snorkeling and dives in the Florida Keys as I researched this book, I was never fortunate to cross paths with any member of the order Octopoda, not in a garden, not in a cave, not on the roof of a submerged Chevy Impala.

So imagine my glee when Kara Norman volunteered her own story about befriending an octopus during a summer stint volunteering in a coral nursery off the shore of St. Thomas in the U.S. Virgin Islands. She and a friend had heard from others that an octopus had been spotted off the beach where they were staying. With masks and snorkels in hand, they rushed out to find it. "We rarely see octopuses in the Florida Keys," she told me. "After a little searching and with a lot of luck, we found the little octopus hiding in the rubble. The little guy had made a nest for shelter." She and her friend floated above for what felt like hours "just watching the beautiful creature." Norman's friend offered the animal a seashell, which it quickly scooped up and wedged into its lair. "We were so happy it accepted her gift!" Norman said. "After that, we visited it every night. On the last night of our trip, we were so sad to say goodbye."

chapter 11
The Long Goodbye

About eighty miles south of New Orleans, as the gull flies, is a narrow spit of land in the Louisiana bayous called Isle de Jean Charles (figure 11.1). You approach the mound on a two-lane causeway called Island Road, which rolls through two miles of open water. On a calm, sunny day, pelicans wheel overhead, and locals congregating on the road's shoulder cast spin lines into the surf—some from the front cabs of their pickups. On a windy day, however, ferocious tides smack the shores and fountains of saltwater leap onto the pavement, flooding the road and making passage impossible for island residents, whose jobs, schools, hospitals, and grocery stores are all on the mainland.[1]

Island Road is sinking fast, and the island is too. This landmass—home to members of two Indigenous communities—has lost about 98 percent of its total area since 1955. That is more than 22,000 acres, or 12,000 soccer fields. If one engages in a conversation with one of the island's remaining eighty-five residents, they will be pointed to the many circles of open water where, a generation or two ago, children ran, horses grazed, and ancestors were buried. We can blame sea level rise, extreme weather, and, most of all, irresponsible practices in water management and oil and gas production for this loss. Today state planners warn that continued sea level rise and coastal erosion will soon render the island uninhabitable.[2]

This goodbye has been a long one. Residents have been filtering away for decades, and leaving with them are the community's cultural practices, healing plants, traditional foods, and lifeways. After nearly twenty years of tribal efforts, in 2016, the U.S. Department of Housing and Urban Development (HUD) awarded the State of Louisiana $48.3 million in Community Development Block

11.1 Isle de Jean Charles in Terrebonne Parish, Louisiana.
(Photo by Rob Verchick)

Grant funds to resettle island communities, as part of the state's application to the 2014 National Disaster Resilience Competition. One of the communities, the Jean Charles Choctaw Nation, worked with the state and other stakeholders to develop the resettlement proposal.

Nevertheless, after years of debate, including over the role (if any) that tribal leadership should play in the decision making, things are falling apart. While the state has acquired new land and has begun building a new development, many Native residents complain the arrangement is unfair and will not suit the tribe's needs. In addition, tension between the state and the tribe is palpable, with tribal representatives accusing state officials of misrepresenting facts and impermissibly diverting funds. Some island residents say they would rather stay put and go down with the ship.[3]

Four thousand miles away, another native community, a Yup'ik village sagging in the marshlands of southwestern Alaska, is also migrating because of climate change. The village, Newtok, has a population of about four hundred. In recent years it has seen an average loss of more than eighty feet of land annually, which has been caused by a combination of river scour, storm surge, and rapidly thawing permafrost induced by climate change. As of October 2019, Newtok had lost its barge landing, its sanitary landfill, its airstrip, and even its source of drinking water. Like the Jean Charles Choctaw Nation, this Alaska Native community had lobbied federal officials for decades for help to relocate to a safer place—in this case, a village nine miles southeast called Mertavik. And like the Louisiana tribe, it sought a major grant under the 2014 National Disaster Resilience Competition. The proposal, however, was rejected in favor of other communities in need. Village leaders then petitioned President Obama to unlock federal emergency funds by declaring a major disaster. President Obama denied this request.[4]

Since that time, village leaders have managed to patch together an array of grants totaling $64 million from federal, state, and private sources to support the first stages of relocation. Housing and infrastructure are being built at Mertavik, and more than a hundred people have moved there full-time. Planners estimate it would take another $115 million to finish the job, but at this rate, Newtok will be rendered uninhabitable before relocation is complete. In addition, tension between the state and the village is palpable, with village representatives accusing state officials of misrepresenting facts and impermissibly diverting funds. Short on funds, the village has resorted to redirecting the bulk of its federal Covid-relief funds to build new houses in Mertavik. By characterizing the new buildings as "isolation homes," the people of Newtok apparently hope this novel use will qualify as pandemic response.[5]

The residents of Newtok and Isle de Jean Charles are not alone. According to recent estimates, millions of U.S. residents are in danger of being displaced by sea level rise and other climate disruptions before the end of the century. Several coastal communities in isolated areas around the country are considering relocation because of sea level rise and erosion. A few are actively planning their exit strategy.[6]

How will the nation decide which households or communities get priority? Who will manage and pay for their resettlement? What steps should be taken now? As the Biden administration labors to tame the climate crisis, the challenge of domestic migration looms large.

<hr />

The trouble is that there are *a lot* of dangerous places. Think back to the classroom exercise, "Escape from Climate Change" that I wrote about chapter 1. Is anywhere *really* safe? Not coastal Louisiana or Florida or Virginia. Not Rapid City, Idaho. And, no, not coastal Alaska.

As we learned in chapter 4, climate breakdown punishes the powerless first. While New Yorkers consider designs for a multibillion-dollar seawall to protect Wall Street, working-class folks have fewer options in Alaska's marshlands or the Gulf Coast swamps—let alone in Hazira, India.[7] Already, global warming is shaping a "climate underclass" whose property, culture, and well-being are under threat. Building climate resilience and managing voluntary retreat is more than a smart choice: it is a moral duty.

As of 2017, at least seventeen U.S. communities have already started a process of managed retreat on account of climate change.[8] Yet the federal government remains grievously unprepared to handle the challenge of climate migration. There is no dedicated funding, no lead agency, and no recommended framework for guiding communities on such a tortuous journey.

Even so, the U.S. Global Climate Research Program continues to warn of the potential need to relocate millions of people and billions of dollars' worth of infrastructure, portending legal, financial, and equity issues that officials have little idea how to address. In its many reports on climate impacts, the U.S. Government Accountability Office (GAO) has become increasingly alarmed about federal fiscal exposure, noting that since 2005, federal funding for disaster assistance has totaled at least $460 billion. The GAO continues to insist the government needs a robust climate migration program built to scale.[9]

In its final two years, the Obama administration took several actions to guide federal officials in climate relocation efforts, most notably in Alaska, where the imminent need is greatest. Following a visit to the Arctic Circle to review coastal damage, President Obama directed the Denali Commission, an independent federal agency charged with supporting Alaska's rural communities, to play a lead coordination role in assisting relocation efforts. In his last month in office, President Obama established an interagency working group on voluntary relocation to "develop a framework for managed retreat."[10]

Then President Trump was elected. President Obama's executive orders on climate change were rescinded, funding for the Denali Commission was whittled down, and the relocation working group was smashed like crusted snow under an ice cleat.

As I write, President Biden seems eager to dislodge the many cleats that have mangled climate policy in the last four years. Through the new White House Office of Domestic Climate Policy, his administration has elevated climate change issues throughout the executive branch. In January 2021, President

Biden issued a directive titled "Executive Order on Tackling the Climate Crisis at Home and Abroad," which established a National Climate Task Force aimed at reducing greenhouse gases, preparing for climate change impacts, and ensuring a just economic transition to sources of renewable energy like wind and solar. The order, which does not specifically address domestic migration or relocation, does stress the need for building "sustainable infrastructure" and preparing for climate change impacts "across rural, urban, and Tribal areas," suggesting a scope large enough to encompass resettlement.[11] This is hopeful news, because it is imperative that federal policies reach beyond the ad hoc actions of the past to embrace a permanent framework that is up to the task.

The truth is, there are hundreds, maybe thousands, of communities across America that will flood, burn, or blow away by the end of the century, irrespective of how much carbon dioxide we are able to cut. The United States needs a program that coordinates across sectors, operates at scale, and emphasizes the needs of the most vulnerable populations.

Before moving forward, it might be helpful to make some conceptual distinctions. As a concept, climate-induced migration is incredibly broad. Leaving aside the point that migrations almost always have more than one root, households move in different ways. The Indigenous peoples of Isle de Jean Charles and Newtok are engaged in what we might call "resettlement." This is the relocation of a whole people or community. The process is complicated and expensive, requiring years of planning, the building of new infrastructure, and a high level of collective decision making.

Most household migrations are performed individually and follow a pattern of gradually embraced commitment and planning. The drivers of this type of move are often expressed as concerns about the local economy, the cost of housing (including insurance), personal safety, or other considerations. "Climate change" might not be written on anyone's pro/con list, but you can see how impacts like hurricanes or heat waves might factor in.

Other relocations, like the diasporas that followed Hurricane Maria in Puerto Rico in 2017 or the Dust Bowl drought of the 1930s, are more sudden and surprising. These events, which I think of as "displacements," are particularly hard on groups that lack economic security.

By the way, let's not refer to any of the people discussed in this chapter as "climate refugees," as news headlines occasionally do. Legally speaking, "refugees" describe persons who are leaving one country for another. By a wide margin, most people migrating for reasons related to climate impacts will do so inside the country in which they already reside. What's more, the term "refugee" is now so freighted with negativity that some people consider it pejorative. Whatever your view on that subject, it seems likely that talk of "climate refugees" is not going to enlighten our discussion.[12]

Choosing a home is an inherently stressful process. There are countless factors to consider, from renovations and interior design to location and market fluctuations and of course, mortgage and insurance rates. It's an investment, but if you're smart, a lucrative one. Most people buy where they feel the value will appreciate over time—an up-and-coming neighborhood in a popular city, for instance, or some place with a view that only gets better.

But many places that would have been desirable in the past are now experiencing the impacts of the climate crisis. Think about a place like Brooklyn, New York's hippest borough and an undeniably expensive one, where the streets flood in severe storms and subways fill with water. In September 2021, Hurricane Ida dropped as much rain in Manhattan's Central Park in an hour (3.15 inches) as Chicago typically gets the whole month of September. More than fifty people died. Storms are getting stronger and more spontaneous, an effect referred to as "rapid intensification." Cities don't have enough time to prepare. Rapid intensification also makes it harder to evaluate storm risk. Meteorologists may be able to predict a nor'easter's path, but not its strength.

Then there's Miami, the epitome of luxury and glamor, where many of its neighborhoods are flooded even amid sunshine. Cities like Seattle and Portland, Oregon, which rarely saw extreme heat in the past, now prepare for summer days topping 100 degrees F. In California, thousands of homes and buildings are damaged or destroyed by annual wildfires.

In terms of real estate, damages from these natural disasters are hitting record highs. Floods, hurricanes, mudslides, and wildfires cost hundreds of billions in damage in the United States. Accordingly, some investment companies have begun planning for climate risks and trying to tamp down the costs. Strategies include diversifying their portfolios, mapping risk, armoring properties, and pushing for more government safeguards with policymakers. The private sector is lending a hand too.

You may remember Redfin from chapter 7. That's the online real-estate brokerage firm that reported higher rates of appreciation for West Coast residences in places where wildfire risk is low. Redfin is a giant in the real estate market. In 2020 it helped customers buy or sell more than 310,000 homes, a value worth more than $152 billion. That year, it began grading its listings for climate risk, giving each home a ranking based on scores for things like fire, flood, storm, drought, and heat. Redfin's scores are calculated using advanced technology and analytical tools. If an area affected by one of these risks has taken steps to manage the hazards, it will score better.

"I remember I was looking at Phoenix, Arizona," Daryl Fairweather, Redfin's chief economist, told me, "and I was expecting to see a really high number. But because Phoenix has been dealing with drought for a very long time, they already have a lot of the infrastructure in place to deal with that. For example, people don't have green lawns in Phoenix anymore, and they're already doing water rationing there. But if you look just over to Maricopa," she said, referring to an outlying city thirty miles away, "it does have high drought risk because it doesn't have that same resiliency."

The real estate site also researches trends in migration to analyze where people are moving to and why they're leaving. According to Fairweather, people are relocating to areas with the highest climate risk and leaving those with less threat. That's exactly the *opposite* of what should be happening. "People are moving to places like Arizona and Colorado and Florida," she told me. "Miami was the number one migration destination of any Metro area. And Miami is on the coast. A lot of those properties have very high flood risk." As a result, she expects to see more and more displacement in the Sunshine State.

The octopus in the parking garage might be Florida's most beguiling flood victim, but it will not be the only one. Studies show that a quarter of all people living in the state risk some form of flooding—that's 3.5 million people on the coast and 1.5 million people living inland. To learn how that fact is affecting Florida's housing market, I contacted Philip Mulder, an economist at the University of Pennsylvania's Wharton School of Business. He studies the effects of sea level rise in real estate markets. What Mulder found in his study of Florida, he said, is a little strange: flood risk does not appear to be affecting sale prices, but, in some circumstances, it is affecting sales volume. From 2001 to 2013, he told me, sale prices and sales volume tracked each other closely. In that period, both rose during the housing boom and tanked during the housing bust.

"But then in 2013, something suddenly changed," he said. "We saw that home sale volume started to decline in those markets that would basically be almost completely underwater with six feet of sea level rise." What was special about 2013? That was the year the UN Intergovernmental Panel on Climate Change issued its fifth assessment report indicating that sea level projections were even worse than we thought. In Florida, Google searches for "sea level rise" spiked during that time.

What was odd was that the scary headlines dampened volume but not price. "Even as home sale volume started to plummet," Mulder explained, "home prices continued to grow as if nothing was going on." At first, he thought lenders might be tightening credit in areas exposed to rising seas. But the evidence didn't bear that out. It turns out, he said, that the sellers were responsible for the mismatch. They were still holding out for top dollar—the climate crisis be damned. It took three more years, until 2016, before the rate of growth on sale prices started to decline. Note that I said "the rate of growth." The prices themselves are still climbing, though at a slower pace, keep in mind that we're talking about coast-line villas that in fifty years could plausibly be underwater.

I take two messages from this story. First, markets alone—even relatively informed ones—are not going to move people from danger fast enough. It takes time for supply and demand to negotiate terms. And when people's homes and communities are at stake, market forces are only part of the calculus. There are families, neighbors, schools, jobs, places of worship, and so much else to con-sider. For most people living full lives, goodbyes are long, not short.

The second message is that the phenomenon we've been discussing is packed with a lot of privilege. I recall reading a news article about a homeowner in Miami whose garage flooded every single day because of the local tides. He had already spent $250,000 to "floodproof" the house. When asked how he justified the investment, he said could never leave—the view was just too beautiful.[13] Homeowners in wealthier communities, according to Fairweather, are also more likely to persuade their municipalities to pay for collective protections like sand dunes and seawalls.

This raises the issue of climate and caste. Income, race, gender, LGBTQ sta-tus, and other characteristics all affect people's access to housing. And, as Fair-weather told me, "Housing policy is 100 percent climate policy." Municipalities should be working not only to build more affordable housing but also put it in climate-safe places. Instead, according to Fairweather, "We've built fewer homes in the last decade than we have in any decade going back to the 1960s." A resilient home in a safe location, she suggested, might someday become a "luxury good," available only to a certain class. That's unfair—and dangerous. "If somebody wants to leave a climate-risky area, coastal Louisiana, for example, to move to Wisconsin or Minnesota, I think that we should be facilitating that," she said. "There's going to be some people who want to stay behind if they have the money, and that's fine. But for people who don't have the money, they may need some intervention in some cases."

Some coastal property owners will rely on private or public insurance as a form of resilience, allowing owners to recoup losses and build back better, if they choose. But there are few private flood insurance policies around. Mortgage holders located within the federally designated flood zones are required to buy flood insurance, which is almost always obtained through the National Flood Insurance Program (NFIP), which was initiated in 1968 and is run by the Federal Emergency Management Agency (FEMA). Most flood-prone properties built before the advent of federal flood maps and the NFIP enjoy subsidized, nonmarket-based premiums. Today, roughly a quarter of all properties insured under the program receive such a subsidy.[14] This arrangement seemed fair to lawmakers fifty years ago. The idea was that local governments would steer development away from these unsafe areas and use their regulatory powers to restrict land use and strengthen building codes. That didn't happen as much as it should have.

It was a problem of mismatched control centers, a contrast to the multiple-brained octopus we studied in the last chapter. At one level of control, local governments took on the politically unpopular job of tightening zoning and safety laws. At another, the federal government promised to subsidize the flood insurance of residents without sufficient systems of oversight and accountability. The munificent, hub-based brain had too little control over what the waggling arms were doing. Today, in the wake of more catastrophic storms like Katrina, Rita, and Sandy, the NFIP is more than $20 billion in the hole.[15]

In response, Congress has in recent years tried to dial back some of those subsidies, particularly for homes with repetitive claims. In states like Texas, New Jersey, and Virginia, many of those homes are becoming unsellable. That effect falls especially hard on lower-income families whose sole or primary source of wealth may be a house no one can insure.[16] Reducing insurance subsidies may also have a racialized effect, since discrimination in the housing market has historically limited families of color in obtaining safer housing. The tentacles of caste reach everywhere. For years, lawmakers have considered workarounds, from needs-based insurance vouchers to voluntary buyouts. But because doing anything at scale is expensive and politically divisive, Congress seems content to kick the can down the road and allow the NFIP to rack up insurmountable debt.

Government must play a more active role in ensuring safe, affordable housing for everyone. That may mean allowing denser development in safe places, boosting land-use restrictions in unsafe areas, and strengthening building codes. Where these policies increase the cost of housing, need-based subsidies should be made available. As for flood insurance, while it makes sense that premiums should rise with risk, the government has a duty to see that those with lower incomes are buffered from financial ruin.

In addition, the private sector has an important role to play in providing climate-risk information to help people make better decisions when choosing a home. The demand is there. According to surveys conducted by Redfin in 2020, 49 percent of Americans who planned to move in the next year said that natural disasters or extreme temperatures factored into their decision to relocate. In addition, said Fairweather, "we found that three-quarters of Americans are hesitant to buy homes in areas with climate risk, and that one in five Americans believe climate change is already negatively impacting home values in their area."

So, let's get this straight: Home buyers are concerned about climate change and the effect it has on property values. But when you look at where people are moving, they're still going in the riskiest places. Climate is a factor, but not *the* factor. Not yet. Which is something I understand. I live most of the year in New Orleans, after all. I love the place. Thus, I was curious to see how my neighborhood compared to other neighborhoods in other cities on Redfin's database. I asked Fairweather to look up my neighborhood and compare it to her neighborhood in Williams Bay, Wisconsin. After a few clicks on her laptop, she had the answer.

On a risk scale of 1 to 100 (where 100 is really bad), Fairweather's neighborhood scored 63 on storm risk, 75 on heat risk, 24 on drought risk, and 20 on wildfire risk. Her flood risk (which is evaluated on a different scale) was "low." Fairweather doesn't like her storm risk. "It's probably the most troubling," she said. "But there are things that I can do, like making sure the storm water flows away from my house and doesn't flood my basement. I can make sure my roof is in good shape." As for the heat risk, "you can prepare for that with air conditioning," she said. "A lot of this comes down to like what risk you personally can tolerate." Fairweather said she likes where she lives and that she and her family have no plans to move.

As for my neighborhood in New Orleans—the one that was swamped after Hurricane Katrina: I have to say I was pleasantly surprised. My neighborhood scored 70 on storm risk (natch), a 57 on heat risk, an 8 on drought risk, and 0 on wildfire risk. My flood risk was "moderate." You won't see this on Redfin, but I can also walk to Mardi Gras parades and bike to Jazz Fest. So, like Fairweather, I have decided to happily stay put.

We've learned that climate impacts are affecting the security and livability of many places in the United States and elsewhere. Many communities and their residents are already preparing for these changes. Some residents, however, may decide to relocate to another place. A key theme here is autonomy, or choice. Some people have the choice to decide whether or where to move. Many don't. Some may have lost homes to extreme weather events and can't afford to rebuild. Having been displaced, they seek to move. If they don't have the means to move, they stay and muddle on as best they can.

The problem, according to Rachel Cleetus, a policy director at the Union for Concerned Scientists, is that our political leaders are not treating climate disruptions as part of a pattern. "We're still reacting to these climate-related impacts as if they're one-off disasters—a hurricane, a wildfire, a flood—when in fact what we're seeing is a worsening trend," she told me. In addition, "There's been recent research showing that a lot of this disaster aid goes to whiter, wealthier communities, and that Black and Brown communities have been at a disadvantage in accessing this disaster aid."

Large-scale relocations are not new in U.S. history. In the Great Northward Migration, six million African Americans moved from the rural southern United States to the urban Northeast, Midwest, and West between 1916 and 1970. It was fueled by poor economic conditions, natural disasters like the 1927 Mississippi River flood, and racial discrimination. The Dust Bowl exodus was the largest relocation in American history. By 1940, 2.5 million people had moved out of the Plains states, including 200,000 who settled in California. Those were reactions to immense cultural, economic, and environmental forces. Neither event was officially planned, of course, and both caused enormous disruptions to economic markets and local systems of governance. Less dramatically, we've seen migrations toward the Sunbelt and from small towns to urban and suburban centers.

And because this is so normal, we may not have noticed that some recent passages could have been prompted by the climate crisis. Consider those crossing the border from Mexico into the United States. Almost 1.7 million migrants were apprehended at the border last year, with research suggesting fragile ecosystems and environmental upheaval provided stressors to weak economies and poor state institutions.[17]

But even if a drought in Mexico put a family at risk, spurring them to migrate to the United States, that family would have no case at the border. Immigrants must provide cause for refugee status, and officials look only for political persecution, not climate-induced threats. In 2021, a government task force focusing on climate and migration recommended closer monitoring of people forced to leave their homes because of natural disasters and began working with Congress on a plan to add droughts, floods, and wildfires and other climate-related reasons to be considered in granting refugee status.[18] But we're not there yet.

And let's say a migrant family manages to settle somewhere like California, which has weather and terrain they are familiar with. They might look for jobs in the agricultural sector and establish a new home. Life might get better for a time. But California is getting hotter and drier too. Crop yields decrease. Drought destroys harvests. Smoke from wildfires pollutes the air. Communities that already lacked stability in terms of employment and social services will face renewed hardship.

It's not just California either. The village of Hatch, New Mexico, is known for producing chiles. The town hosts an annual Hatch Chile Festival that attracts thousands of visitors from around the country. Like the almond industry in California however, farmers in New Mexico worry about the crop's future profitability. Continued drought and an unprecedented workforce shortage worsened by the pandemic have hurt the agribusiness. Chiles won't fruit above 95 degrees. Rising temperatures and lack of water upended the lives of many migrant workers who had moved to Hatch from their previous locations. They were forced to move again. When farming became prosperous again, New Mexico was faced with a labor shortage. Farms offered to raise wages, but uncertainty about the climate kept migrant workers away.

One lesson I take from this is that retreat is not always the best answer. Perhaps the chile farms in New Mexico could *adjust* instead by experimenting with new varieties of seed, investing in innovative irrigation techniques, or reallocating water rights. Ilan Kelman, an expert in climate migration at University College London, told me that the real point is not whether people should stay or go, but about what services and what choices they should have in the situation they are in. He reminded me that sometimes those who flee are the lucky ones.

"When there is a problem and people feel that they are forced to migrate," he told me, "it's the people who have the most resources who are able to get out first." He raised the example of Afghanistan. "A few days before the Taliban moved into Kabul, the airport was packed with people who could immediately afford to buy any plane ticket to go anywhere. After that, you had the middle classes or the middle lower classes who may have had to sell everything to get out. But they had the option, and they did that. The ones left behind have no cash and nothing to sell." So it is with global warming. "As climate change impacts local environments," he said, "those who cannot adjust, cannot leave, cannot change—those who have nothing to begin with—end up being trapped and hurt the most."

How might we, as a society, build a better template for responding to the needs of climate migrants—a template that is flexible, forward-looking, fit-to-scale, and fair? As a legal scholar, I've thought about this a lot. I have a set of staged recommendations that focus on assigning leadership, assessing community risk, inventorying the financial and legal resources, and expanding authority. In terms of political possibility, the first three fall into the "easy to somewhat challenging" category. The fourth one is harder.

The first thing I would recommend would be to assign a leader at the federal level. In its extensive 2020 report on climate relocation, having combed the published research and interviewed a range of stakeholders, the GAO announced what it determined to be "the key challenge to climate migration as a resilience strategy," namely "unclear federal leadership."

The GAO noted, for instance, that while the Denali Commission was able to play a significant role in Newtok's relocation, the commission had no authority to assist other Alaska Native villages in navigating federal program requirements or securing funding and technical assistance. Unclear leadership, according to the GAO, also hobbled and prolonged the twenty-year resettlement process for Isle de Jean Charles. To raise just one example, because neither of the tribes present on the island was federally recognized, HUD was required to interact with state rather than tribal officials. The arrangement led to confusion, distrust, and allegations of bad faith among the parties.[19]

The president could assign immediate information-gathering projects to an entity that already exists, such as the National Oceanic and Atmospheric Administration or the U.S. Environmental Protection Agency. With some leadership in place, the government's next move should be to comprehensively assess the scale of the country's potential climate migration. As the federal government noted in its fourth climate assessment, there exists only vague ideas about the magnitude, timing, and spatial distribution of this problem.[20] Some of that uncertainty involves geophysical factors like the variability of Midwestern downpours or the fire sensitivity of emaciated forests. The GAO reports that today's researchers are nowhere near having models with the resolution necessary to identify the hundreds or thousands of places that could face an existential threat.

In addition to the geophysical uncertainties, there are social, economic, and cultural questions. Whether a family or community chooses migration will depend on its access to relocation resources, its cultural ties to a place, and, of course, the other options available. Nobody expects the well-heeled residents of Lower Manhattan to pick up and leave; the U.S. Army Corps of Engineers is already looking into a multibillion-dollar seawall. Conversely, many residents in Louisiana's coastal parishes are planning moves.

Parish communities lost the fight for a regional levee system because of economic and environmental concerns.[21] Louisiana's coastal adaptation plan (significantly aided by HUD's Community Development Block Grant Disaster Recovery Program) is offering millions of dollars in risk-mitigation grants for "floodproofing" homes within fourteen feet of FEMA's projected base flood elevation. The hundreds of homes that do *not* meet this criterion will be invited to seek voluntary buyouts, whisking them into a new stream of climate migration.

One point is that the need for resettlement is determined by more than geographic location; it is also determined by the economic and political power a climate-affected group can muster to protect itself in other ways. A second point is that even where retreat *is* a group's preferred option, social and economic factors may inhibit its ability to exercise it. A poor family or community that lacks the resources to fortify itself against a flood-prone river will probably also lack the resources to acquire safer land and resettle it.

Institutional barriers can also disproportionately affect socially vulnerable populations from choosing adaptive options, including relocation. For instance, a town completely dependent on recovery funds from FEMA to rebuild damaged property might find that the costs of moving to another site are not covered. To take another example, the GAO reports that some federal programs may "constrain tribal communities' ability to pursue self-determined management of their resources and built environment" because they do not "account for the unique context of tribal communities and tribal sovereignty."

A second inventory involving funding opportunities and legal authorities is also needed. Federal programs fall dramatically short in supporting climate adaptation efforts generally and migration efforts in particular.[22] Most adaptation resources come through ad hoc agency projects that are driven by primary missions (i.e., secure housing, sound transportation) that, while relevant to climate impacts, exist independent of them.

Plus, the programs that individuals rely on to help communities replace damaged infrastructure and housing after disasters were never intended to account for the size and complexity of relocating entire populations. The fragmented and ad hoc nature of hazard mitigation funding has made it difficult to even know how much the federal government is spending on this work. State funding is similarly difficult to track.[23]

Legal authority is, of course, baked into the soufflé of all funding options. An agency's authority is often limited to situations where disaster has already hit. Sometimes, though, there are ways around this problem in the form of "pre-disaster" aid. In addition, there are legal authorities that might help ensure that migration and other forms of resilience aid go to the socially vulnerable populations who need it most—or at least do not make things worse for such groups. Experts do not often think of laws like these as part of "disaster law," but they should.[24]

My last recommendation is the most difficult to achieve politically because it would require legislative action from a Congress that is notoriously divided over climate change.[25] The truth is that, even with sterling executive leadership and the most robust informational inventories, the government cannot serve an effective climate relocation program—one that makes good on its moral obligation to the citizenry and reduces the treasury's astronomical fiscal exposure—without bold and creative federal legislation.

The most ambitious plan would empower and fund an existing or new federal agency to oversee and advise voluntary, community-led climate relocations. One idea might be to implement on a national level the vision President Obama had for the Denali Commission in the latter half of his second term. This new agency would function as a counselor to eligible communities, with formal relationships with all relevant federal agencies to provide a kind of one-stop shop for technical assistance, bureaucratic troubleshooting, the assembling of grant packages, and social and cultural needs.

Along with new agency leadership would come new funding—funding that is equal to the scale of the problem, based on a predictable source of revenue, earmarked specifically for climate resilience efforts (perhaps, more specifically, relocation efforts) and, crucially, funding that is available *before* disaster strikes. Practically speaking, it probably makes sense for funding sources to be divided among agencies and designed to suit an array of particular uses or recipients. It might also make sense to integrate such funds into the current legislative agenda. New climate legislation could include resilience grants funded by carbon fees, the sale of pollution credits, or other mechanisms. Finally, when Congress gets serious about revamping the National Flood Insurance Program, it should consider a generous package of voluntary buyout provisions—designed to promote collective, community-based relocation decisions—to help populations that choose to move to higher ground.

❦

It's hard to walk the waterlogged fields on Isle de Jean Charles and not feel a sting of despair. There is much to do, after all, and the United States has dismissed many opportunities for early action. But that does not absolve the government and its citizenry from persevering. Tackling the migration challenge requires a deliberate, staged approach that includes establishing a specific and accountable leadership structure, assessing the geophysical and socioeconomic vulnerabilities of communities at risk, inventorying existing relevant laws, and enacting federal legislation that organizes a unified response, fills the gaps in existing authority, and provides perpetual and reliable revenue streams to address the issue at scale. Hope is alive, but time is running short. The water is lapping at our heels.

chapter 12
Persist and Prevail

In July, outside the Paradise Inn in Washington's Mount Rainier National Park (elevation 5,400 feet), I still need a fleece to keep warm. In the shadow of the park's snow-covered volcano, I stand surrounded by bird chatter and meadows sparkling with wildflowers. Built in 1916, the inn is one of those grand, rustic affairs, reminiscent of Yellowstone's famous lodge. The surrounding swales and footpaths feel almost like a sprawling urban park, which makes sense once you realize these grounds were designed by Frederick Law Olmstead, the architect of New York City's Central Park and Boston's "Emerald Necklace" gardens. I remember a news article from a few years back about how Mount Rainier's iconic bands of wildflowers were slowly retreating to higher elevations away from Paradise Inn. Park scientists attributed this to temperatures caused by climate change. There was some debate, as I recall, about whether park staff should manually seed the meadows where lodge visitors gather or to let the buttercups and salmonberries crawl naturally uphill. I don't know where they ended up on that.

Today I'm not here for the wildflowers, or the bird chatter. I've come to hike what's called the "Skyline Trail," a course that takes you upward through Hobbit forests and past mountain vistas to a snowy patch from which you have a clear sighting, though at some distance, of the mighty Nisqually Glacier. Mount Rainier, whose summit rises to 14,411 feet above sea level, is home to twenty-seven major glaciers. Together, they amount to about one cubic mile of ice and snow. The glaciers attract mountain climbers from around the world, some of whom use Mount Rainier as training for future expeditions on Mount Denali, in Alaska, or Mount Everest. Nisqually is the longest and most accessible glacier in the park.

For many years, I've taken an interest in these glacier systems. Back in 1999, my wife, Heidi, and I climbed Mount Rainier to mark our tenth wedding anniversary. Carrying fifty-pound packs and sometimes roped to fellow climbers, we followed a route that took us through the Muir Snowfield (which parallels the Nisqually Glacier) and then onto parts of three other glaciers—the Cowlitz, the Ingraham, and the Emmons. Since then, I've visited Mount Rainier every year or so, usually with my sons, hiking the Skyline Trail. We'd marvel at the subalpine firs, straight as bottlebrushes, and I'd offer a quarter to anyone who could more specifically name a tree. We'd imitate the ravens croaking overhead and giggle at the hamster-like pikas playing hide-and-seek in the rocks. Then we would arrive at the lookout point for Nisqually Glacier. No more croaking. No more giggling. All we could do is gape and grin.

Today, from my vantage on the Skyline Trail, the Nisqually seems to have lost a little weight since my last visit. It's still a colossus—don't get me wrong—and a thing of beauty. The ice reminds me of a mass of veined Italian marble, its chiseled braids shaded in rose, orchid, and gunmetal gray. Years ago, when I first started coming up here, I often had trouble distinguishing glacial ice from pockmarked snowfields. A trick I learned was that, regardless of color, *ice* had a certain tenseness, a musculature, that snow powder lacked. Here, the comparison to marble helps again: for like the naked abs of, say, Michelangelo's *David*, the coiled torso of the ancient Nisqually is, beneath the skin, strained and shivering—stretching imperceptibly toward release. Like all of Mount Rainier's glaciers, the Nisqually is "retreating," which is to say melting and draining away.

I take out my binoculars and focus on the bottom end of the glacier, called the terminus. I can just make out a sliver-moon opening and a cataract of milky water tumbling down. On a human scale, I know that opening is ridiculously large, the size of a concert hall or an airplane hangar. That water racing out is filled with tons of sediment, rock, and boulders. There's nothing unusual, in itself, about a glacier spitting out ice water. That's what they do. Over millions of years, an ice sheet grows; the glacier's enormous weight creates pressure and heat inside; internal layers of ice melt, escape as water, and eventually fill downstream rivers and lakes. In a healthy glacier, the ice growth outpaces the water loss. But today, in Mount Rainier National Park, the situation is reversed: less ice is piling up and more water is flowing out.

That excess flow contributes in various ways to downstream flood problems in communities within the park that are located along the rivers that the glacier feeds. Every year or so, it seems, I run into a scouting troop or volunteer service organization of some kind in this park helping to shore up washed-out trails or rehabilitate flooded campgrounds at the lower elevations. They might not see themselves this way, but to me they are like the Louisiana activist Sharon Lavigne in chapter 4 or diving educator Kama Cannon in chapter 10—soldiers for resilience on the front lines of the climate crisis.

———————

As daunting as climate action sometimes seems, there are advantages in tackling resilience. One is that, as I explained in chapter 5, the values at stake (personal security, property protection, community identity) resonate with groups that might not intuitively embrace the work of emissions reductions. Another is that the menu of topical issues is as large as your imagination. You start by asking what you care about, and then ask how climate breakdown will challenge that thing. Addressing that challenge becomes your work. Some of that work will be top-down, broad-scaled, and policy-driven. A global crisis requires big action by government and coalitions of government. I've done my best in this book to lay out some of the marquee items we should think about in this regard and how we might evaluate them.

Still, much of resilience work is more local and concrete. This is where you can make a difference every day. It's just a matter of assessing your interests and deciding what matters most to you. Maybe, like Minerva Hoyt, you like to garden. Maybe, like Kara Norman, you like to dive and snorkel. Maybe, like Bob Marshall or Shirell Parfait-Dardar, you like to hunt and fish. Maybe, like Sam Perkins, you like hiking and see yourself as a caregiver. Maybe, like Wendy Gao, you like public speaking. Maybe, like Cinthia Zermeño Moore, you want to be the best parent you can be.

As a university professor and a former government official, I'm often asked how one might get started in this business of climate action, and, in particular, in building climate resilience. Over time, I've come up with a system I call "LTD": Learn. Talk. Do.

Learn means to study sources of information on your topic of interest. But it also means learning about the values and interests of others around you and, of course, learning about yourself.

Talk means communicating about what you have learned and why you care about what you have learned. As climate scientist Katharine Hayhoe explains, "The most important thing every single one of us can do about climate change is talk about it—why it matters, and how we can fix it—and use our voices to advocate for change within our spheres of influence."[1]

Do is the part where you take action. But you don't have to make it complicated. You can choose a topic to work on and later find it no longer inspires you. Fine, choose something else. It doesn't have to be your career, although it can be. It just has to be something you do regularly, something that becomes part of who you are. That is what Kama Cannon would call a layer of identity. When you see yourself as someone who learns, who talks, who does, that is who you become. And when you start to feel nervous, or despondent, or worse, remember another great piece of advice from Hayhoe: "The antidote to anxiety is action."[2]

We can't pretend that reducing greenhouse gas emissions will get us anywhere near what we need to thrive; we need to march toward resilience. The journey will not be easy and certainly not without pain. But with clear eyes and open hearts we can persist and, just maybe, prevail.

Acknowledgments

Desmond Tutu said that in life the path of joy is connection with others. So it is with writing a book. In this endeavor, a lot of joy and labor was shared.

The first thank-you goes to the many people I got to know and interview about their experience, advocacy, or research concerning climate change. It's impossible to talk with people like Sharon Lavigne and Kara Norman (to name only two) without feeling a little more inspired and hopeful about the hard challenges we face.

I also want to thank Richard Conlin, resident of the Mirador 1000 condominium complex, whose photo and videos of an octopus in his parking garage lit up social media and inspired the title of this book. My friend Dan Farber of the University of California, Berkeley, is the one who first brought that octopus to my attention. I am eternally grateful for his suggestion that we coauthor an op-ed for the *Miami Herald* about climate resilience, which we titled "About that Octopus in the Parking Garage." I must admit the idea for the title was his. With Dan's encouragement, I am surfing that wave still.

I owe much gratitude to Loyola University New Orleans, my professional home since the fall of 2004 (nine months before Hurricane Katrina). Loyola has long supported my teaching and research in the developing fields of disaster and climate law. In particular, my law school dean, Madeleine Landrieu, allowed me the resources and latitude needed to take on such a sprawling topic in book form. In addition to Loyola, I received critical financial support from the Louisiana Board of Regents Endowed Chairs for Eminent Scholars Program, the U.S. Fulbright Scholar Program (for my research in India), and the Australian Research Council (for part of my research on resilient power grids).

My agent, William Clark, embraced this project from the outset and helped me navigate the Covid-era world of publishing with patience and finesse. Above all, his X-ray vision for narrative structure helped me see the core story I was telling so that I could communicate it effectively to a popular audience.

Some narrative features of this book were first developed as part of a podcast I host, *Connect the Dots*, which is produced by the Center for Progressive Reform. For their contributions, I thank the show's producer, Courtney Garcia, and its assistant producer, Maggie Dewane.

As an academic, I have given several presentations on the ideas presented in this book and received a tsunami of helpful suggestions and comments. These events include programs organized by the University of California, Berkeley; the University of California, Los Angeles; the Indian Institute of Technology, Mumbai; the University of Florida; the University of Kansas; Loyola University New Orleans; the National University of Singapore; the University of Oxford; Puson National University; Stanford University; the University of Sydney; Temple University; Tulane University; Vanderbilt University; and Yale University. Readers looking for more formal or technical treatment of my resilience research can find that in some of my more recent scholarly publications, including those appearing in the *University of Illinois Law Review*, *Temple Law Review*, *Duke Environmental Law and Policy Forum*, *Frontiers in Climate*, *Environment and Planning*, and *Philosophical Transactions of the Royal Society*.

I prevailed upon friends, colleagues, and others to critique drafts of the manuscript, in whole and in part. Their insights greatly improved the book. For their efforts, I thank Andrea Armstrong, John Blevins, Camille Manning Broome, Logan Atkinson-Burke, Erik Christensen, Kendra Chritz, Robin Kundis Craig, Rob Glicksman, Michael Gerrard, Alice Grainger, Paul Herber, Michele Janin, Lisa Lindstrom, John Lovett, Heidi Molbak, J. B. Ruhl, Barbara Schneider, Jerry Schneider, Kathleen Sealey, A. R. Siders, Karen Sokol, Allison Stevens, Kelly Verchick, and Scott Willmann.

Others endured repeated conversations about my research, teaching me more about relevant topics or helping me to frame and sharpen my ideas. For that I thank Vicki Arroyo, John Barry, Po Bronson, Maxine Burkett, Gilonne d'Origny, Dan Farber, Shalanda Baker, Nina Kazazian, Nancy Levit, Andy Revkin, Jim Salzman, Joel Scheraga, Lisa Grow Sun, and Tegan Wendland.

I sometimes took the liberty to assign draft chapters of this book to my students at Loyola Law and at the Disaster Resilience Academy at Tulane University's

School of Social Work. Their reactions and insights improved my thinking immensely in my writing. I owe Reggie Ferreira, director of the Disaster Resilience Academy, a special debt for opening my eyes to the world of social work as it applies to climate challenges.

I am also thankful to an outstanding team of student research assistants: Melia Cerrato, Gabriela Carballo, Claire Dulle, Gabrielle Evans, Spring Gaines, Alina Gonzalez, Kimberly Fanshier, Angie Madore, and Caitlion O'Neill. Ford Miller, one of Loyola's top librarians, provided exceptional services, using his Jedi skills to detect and lay hold of materials in myriad academic fields. Thanks also to the patient staff at Mojo Coffee on Freret Street in New Orleans, where parts of this book were written.

There would be no book were if not for the willingness of my publisher, Columbia University Press, to wade into an octopus's parking garage with me. My thanks go to the whole team at Columbia and, in particular, to my editor, Miranda Martin, and my copyeditor, Gregory McNamee.

Of course, this entire project would have been impossible without the support and understanding of my loving family. For their abiding encouragement, I thank my wife, Heidi Molbak, and my sons, Reed, Ty, and Luke. If I could, I would wrap my arms around you eight times over.

Notes

1. LET'S TALK ABOUT THE OCTOPUS

1. Richard Conlin, "Our greeting this morning in our garage . . . after Octopus last night! Global warming does not exist!," *Facebook*, November 14, 2016, https://www.facebook.com/richard.conlin.miami/posts/10210742406306092. For more of the story, see Alex Harris, "Octopus in the Parking Garage Is Climate Change's Canary in a Coal Mine," *Miami Herald*, November 18, 2016; Daniel Farber and Robert Verchick, "About That Octopus in the Parking Garage," *Miami Herald*, December 4, 2016.

2. For the Sealey quotation, see Eileen Mignoni, "The Bizarre Case of the Octopus in the Parking Garage," *Yale Climate Connections*, January 31, 2017, https://www.yaleclimate connections.org/2017/01/the-bizarre-case-of-the-octopus-in-the-parking-garage/.

3. Kathleen Sullivan Sealey, Ray King Burch, and P.-M. Binder, *Will Miami Survive: The Dynamic Interplay Between Floods and Finance* (Cham, Switzerland: Springer, 2018).

4. Ian Urbina, "Perils of Climate Change Could Swamp Coastal Real Estate," *New York Times*, November 24, 2016.

5. Tim Padgett, "As Sea Waters Rise, Miami Begins Elevated Development," *American Magazine*, June 16, 2017.

6. "Climate Change: How Do We Know?," National Aeronautics and Space Administration, https://climate.nasa.gov/evidence/.

7. National Oceanic and Atmospheric Administration, Earth System Research Laboratory, "CO_2 at NOAA's Mauna Loa Observatory Reaches New Milestone: Tops 400 ppm," May 10, 2013, https://www.esrl.noaa.gov/gmd/news/7074.html (noting estimated concentration before the Industrial Revolution).

8. "Trends in Atmospheric Carbon Dioxide," National Oceanic and Atmospheric Administration, Earth System Research Laboratory, https://www.esrl.noaa.gov/gmd/ccgg/trends/monthly.html (reporting carbon dioxide concentration as of February 19, 2020).

9. "Climate Scoreboard," *Climate Interactive*, https://www.climateinteractive.org/programs/scoreboard/.

10. Richard Rood, a climate scientist at the University of Michigan, makes the point succinctly: "Earth's climate is changing rapidly. We know this from billions of observations, documented in thousands of journal papers and texts and summarized every few years by the United Nations' Intergovernmental Panel on Climate Change. The primary cause of that change is the release of carbon dioxide from burning coal, oil and natural gas." Rood, "If We Stopped Emitting Greenhouse Gases Right Now, Would We Stop Climate Change?," *The Conversation*, July 7, 2017, https://theconversation.com/if-we-stopped-emitting-greenhouse-gases-right-now-would-we-stop-climate-change-78882. For farmers, see Helena Bottemiller Evich, "How a Closed-Door Meeting Shows Farmers Are Waking Up on Climate Change," *Politico*, December 9, 2019. For legal matters, see, for example, *Juliana v. United States*, 947 F.3d 1159 (9th Cir. 2020); HR 20 December 2019, 2020 (Urgenda/Netherlands): 10, http://blogs2.law.columbia.edu/climate-change-litigation/wp-content/uploads/sites/16/non-us-case-documents/2020/20200113_2015-HAZA-C0900456689_judgment.pdf. For the Pope, see Pope Francis, *Laudato Si'*, encyclical letter, Vatican website, May 24, 2015, http://www.vatican.va/content/francesco/en/encyclicals/documents/papa-francesco_20150524_enciclica-laudato-si.html.

11. Rood, "If We Stopped Emitting Greenhouse Gases Right Now"; James Hansen, Larissa Nazarenko, Reto Reudy, Makiko Sato, Josh Willis, Anthony Del Genio, and Dorothy Koch, "Earth's Energy Imbalance: Confirmation and Implications," *Science* 301, no. 5727 (2005): 1431–35.

12. Camille St. Onge, "Water Supply Impacts," Washington State Department of Ecology, https://ecology.wa.gov/Air-Climate/Climate-change/Climate-change-the-environment/Water-supply-impacts.

13. Kathryn Schulz, "The Really Big One," *New Yorker*, July 20, 2015.

14. Maria Temming, "The U.S. Power Grid Desperately Needs Upgrades to Handle Climate Change," *Science News*, February 12, 2020, https://www.sciencenews.org/article/u-s-power-grid-desperately-needs-upgrades-handle-climate-change; Caitlyn Kennedy, "Risk of Very Large Fires Could Increase Sixfold by Mid-Century in the U.S.," Climate.gov, August 26, 2015, https://www.climate.gov/news-features/featured-images/risk-very-large-fires-could-increase-sixfold-mid-century-us.

15. Jasmin Fox-Skelly, "There Are Diseases Hidden in Ice, and They Are Waking Up," BBC, May 4, 2017, http://www.bbc.com/earth/story/20170504-there-are-diseases-hidden-in-ice-and-they-are-waking-up.

16. Mathew E. Hauer, "Migration Induced by Sea-Level Rise Could Reshape the US Population Landscape," *Nature Climate Change* 7, no. 5 (2017): 3, https://www.nature.com/articles/nclimate3271.

17. Dan Hernandez, "The Hellish Future of Las Vegas in the Climate Crisis," *The Guardian*, September 3, 2019.

18. "Tampa, Orlando Among Fastest Growing Cities in U.S.," Fox 13 Tampa Bay, April 18, 2019, https://www.fox13news.com/news/tampa-orlando-among-fastest-growing-cities-in-us.

19. Kendra Pierre-Louis, "Want to Escape Global Warming? These Cities Promise Cool Relief," *New York Times*, April 15, 2019.

20. Al Gore, *Earth in the Balance: Ecology and the Human Spirit* (New York: Houghton Mifflin, 1992), 240.

21. Al Gore, *The Future: Six Drivers of Global Change* (New York: Random House, 2013), 303.

22. Rob Law, "Adaptation Is the Poor Cousin of Climate Change Policy," *The Guardian*, April 9, 2019.

23. "Paris Agreement," *United Nations Framework Convention on Climate Change* (December 2015): 5, https://unfccc.int/files/meetings/paris_nov_2015/application/pdf/paris_agreement _english_.pdf.

24. UNFCCC secretariat, "NDC Registry," United Nations Framework Convention on Climate Change, https://www4.unfccc.int/sites/NDCStaging/Pages/All.aspx.

25. Office of the Press Secretary, "President Obama's Plan to Cut Carbon Pollution: Taking Action for Our Kids," The White House, June 25, 2013, https://obamawhitehouse.archives .gov/the-press-office/2013/06/25/fact-sheet-president-obama-s-climate-action-plan.

26. Sarah Kaplan and Andrew Ba Tran, "Nearly 1 in 3 Americans Experienced a Weather Disaster This Summer," *Washington Post*, September 4, 2021.

27. Benji Jones, "How Jackson, Mississippi, Ran Out of Water: A Crisis That's Left Thousands of Residents with No Running Water Was Decades in the Making," *Vox*, September 1, 2022, https://www.vox.com/2022/8/31/23329604/jackson-mississippi-water-crisis.

28. Spencer Bokat-Lindell, "Climate Disaster Is the New Normal. Can We Save Ourselves?," *New York Times*, September 7, 2021.

29. Elizabeth Kolbert, *The Sixth Extinction: An Unnatural History* (New York: Henry Holt, 2014), 81.

30. Craig Welch, "What Animals Are Likely to Go Extinct First Due to Climate Change," *National Geographic,* April 30, 2015 (discussing species overall); Robert R. M. Verchick, "Can 'Loss and Damage' Carry the Load?," *Philosophical Transactions of the Royal Society A: Mathematical, Physical and Engineering Sciences* 376, no. 2119 (April 2, 2018): 7, https://doi .org/10.1098/rsta.2017.0070 (discussing coral).

31. Brian Resnick, "An Australian Ecologist Explains Just How Bad the Fires Are for Wildlife," *Vox*, January 10, 2020, https://www.vox.com/energy-and-environment/2020/1/9 /21057375/australia-fire-wildlife-extinctions-ecology; Karin Brulliard, "A Billion Animals Have Been Caught in Australia's Fires. Some May Go Extinct," *Washington Post*, January 9, 2020; Dyani Lewis, " 'Deathly Silent': Ecologist Describes Australian Wildfires' Devastating Aftermath," *Nature*, January 10, 2020, https://www.nature.com/articles/d41586-020 -00043-2.

32. Gore, *The Future*, 303.

33. Center for Progressive Reform, *From Surviving to Thriving: Equity in Disaster Planning and Recovery* (September 2018) 11, http://www.progressivereform.org/survivingthriving_main .cfm (referring to studies from Paul Hogan, *Rebuilding Infrastructure: The Need for Sustainable and Resilient Solutions*) (2018): 5, https://www.zurichna.com/-/media/project/zwp/zna

/docs/kh/infrastructure/zurich_resilient-infrastructure_final.pdf; National Institute of Building Sciences, *Natural Hazard Mitigation Saves: 2017 Interim Report* (2017): 1, https://www.fema.gov/media-library-data/1516812817859-9f866330bd6a1a93f54cdc61088f310a/MS2_2017InterimReport.pdf.

2. ADAPT OR DIE

1. Rick Potts, *Humanity's Descent: The Consequences of Ecological Instability* (New York: Avon Books, 1996), 57–78. My discussion of human evolution in this chapter draws often from Potts's remarkable book, as well as from the excellent website on human origins maintained by the Smithsonian's National Museum of Natural History, "What Does It Mean to Be Human?," Smithsonian National Museum of Natural History, https://humanorigins.si.edu. Other sources from which I learned so much and that have informed my thinking include Meave Leakey (with Samira Leakey), *The Sediments Of Time: My Lifelong Search for the Past* (Boston: Mariner Books, 2020); Agustín Fuentes, *The Creative Spark: How Imagination Made Humans Exceptional* (New York: Dutton, 2017); Clive Finlayson, *The Improbable Primate: How Water Shaped Human Evolution* (Oxford: Oxford University Press. 2014); and Renée Hetherington and Robert G. B. Reid, *The Climate Connection: Climate Change and Modern Human Evolution* (Cambridge: Cambridge University Press 2010). I recommend them all.

2. Potts, *Humanity's Descent*, 141–52 (discussing the Olorgesalie basin).

3. "The Adaptable Human," *Nova*, October 25, 2009, https://www.pbs.org/wgbh/nova/article/adaptable-human/ (interviewing Rick Potts). See Potts, *Humanity's Descent*, 235 ("By broadening the response to large, long-term fluctuation, the generalist gains the power to overcome disparities in the specific factors affecting its lineage.")

4. See, e.g., James Borg and Alistair Channon, "Testing the Variability Selection Hypothesis: The Adoption of Social Learning in Increasingly Variable Environments," *Artificial Life* 13 (2012): 317–24.

5. Potts, *Humanity's Descent*, 7 (important capacities); Hetherington and Reid, *The Climate Connection*, 22 ("genetic" and "behavioral" adaptation).

6. Pascale Gerbault, Anke Liebert, Yuval Itan, Adam Powell, Mathias Currat, Joachim Burger, Dallas M. Swallow, and Mark G. Thomas, "Evolution of Lactase Persistence: An Example of Human Niche Construction," *Philosophical Transactions of the Royal Society of London B, Biological Sciences* 366, no. 1566 (2011): 863–77.

7. Guoqiang Xing, Clifford Qualls, Luis Huicho, Maria River-Ch, Tsering Stobdan, Marat Slessarev, Eitan Prisman, et al., "Adaptation and Mal-Adaptation to Ambient Hypoxia; Andean, Ethiopian and Himalayan Patterns," *PLoS One* 3, no. 6 (2008): e2342.

8. Dennis O'Neil, "Early Human Culture," Palomar College, 1999, https://www2.palomar.edu/anthro/homo/homo_4.htm; *Encyclopaedia Britannica* (2015), s.v. "Zhoukoudian," https://www.britannica.com/place/Zhoukoudian.

9. "The Adaptable Human"; see Potts, *Humanity's Descent*, 249 (discussing importance of symbolic thought and cultural institutions in adapting to a changing climate).

10. Dietrich Stout, Erin Hecht, Nada Khreisheh, Bruce Bradley, and Thierry Chaminade, "Cognitive Demands of Lower Paleolithic Toolmaking," *PLoS One* 10, no. 5 (2015): 1, https://doi.org/10.1371/journal.pone.0121804.

11. "What Does It Mean to Be Human? Brains," Smithsonian National Museum of Natural History, https://humanorigins.si.edu/human-characteristics/brains; Potts, *Humanity's Descent*, 126; Hetherington and Reid, *The Climate Connection*, 211–34.

12. "What Does It Mean to Be Human?"

13. Lucia F. Jacobs, "The Navigational Nose: A New Hypothesis for the Function of the Human External Pyramid," *Journal of Experimental Biology* 6, no. 1 (2019): 222, https://journals.biologists.com/jeb/article/222/Suppl_1/jeb186924/2802/The-navigational-nose-a-new-hypothesis-for-the.

14. Pasquale Raia, Alessandro Mondanaro, Marina Melchionna, Silvia Castiglione, Carmela Serio, and Lorenzo Rook, "Past Extinctions of Homo Species Coincided with Increased Vulnerability to Climatic Change," *One Earth*, October 15, 2020, https://www.cell.com/one-earth/fulltext/S2590-3322(20)30476-0. A team of Australian researchers recently suggested that, in addition to climate change, some *Homo erectus* populations may have just gotten too "lazy," choosing to rely on local substandard materials for tool making rather than hiking to a nearby outcropping where stronger stone was available. Australian National University, "Laziness Helped Lead to Extinction of *Homo erectus*," *Science Daily*, August 10, 2018, https://www.sciencedaily.com/releases/2018/08/180810091542.htm.

15. Potts, *Humanity's Descent*, 2.

16. Potts, *Humanity's Descent*, 277–79.

17. Kurt Vonnegut, *Galápagos* (New York: Delacorte Press/Seymour Lawrence, 1985), 189, 270.

3. SPRAWLING BRAINS AND RUBBER ARMS

1. *My Octopus Teacher*, dir. Pippa Ehrlich and James Reed (Netflix, 2020).

2. See the Sea Change Project, the website for *My Octopus Teacher*, https://seachangeproject.com/my-octopus-teacher (quoting reviewer Angie Driscoll).

3. Peter Godfrey-Smith, *Other Minds: The Octopus, the Sea, and the Deep Origins of Consciousness* (New York: Farrar, Straus and Giroux, 2016), 54–59 (octopus talents); Signe Dean, "Octopus and Squid Evolution Is Officially Weirder Than We Could Have Ever Imagined," *Science Alert*, March 6, 2019 (editing genetic information); Shahar Alon, Sandra C. Garrett, Erez Y. Levanon, Sara Olson, Brenton R. Graveley, Joshua J. C. Rosenthal, and Eli Eisenberg, "The Majority of Transcripts in the Squid Nervous System Are Extensively Recoded by A-to-I RNA Editing," *eLife* 4 (2015): e05198, https://doi.org/10.7554/eLife.05198 (same).

4. D. E. Alexander, "Resilience and Disaster Risk Reduction: An Etymological Journey," *Natural Hazard and Earth System Sciences* 13 (2013): 2707, 2709, https://doi.org/10.5194/nhess -13-2707-2013.

5. "Building Your Resilience," American Psychological Association, https://www.apa.org /topics/resilience.

6. C. S. Holling, "Resilience and Stability of Ecological Systems," *Annual Review of Ecology and Systematics* 4 (1973): 1, 17.

7. Holling, "Resilience and Stability of Ecological Systems," 14, 15.

8. Russell R. Dynes, "The Dialogue Between Voltaire and Rousseau on the Lisbon Earthquake: The Emergence of a Social Scientific View," *International Journal of Mass Emergencies and Disasters* 18 (2000): 97, 97 ("It is appropriate to call the Lisbon earthquake the first modern disaster."). Some historians have wondered whether the Lisbon earthquake deserves the credit it often is given for prompting a cultural debate about the origins of natural disasters and other misfortunes. Robert H. Brown, "'The Demonic' Earthquake: Goethe's Myth of the Lisbon Earthquake and Fear of Modern Change," *German Studies Review* 15 (1992): 475, 478 (noting that skepticism toward theodic explanations for disaster predates the Lisbon quake). Still, the symbolic importance of the earthquake in Western thought is "undisputed." Kristian Cederval Lauta, "Exceptions and Norms: The Law on Natural Disasters" (PhD diss., University of Copenhagen, 2012), 43n42; Dynes, "The Dialogue Between Voltaire and Rousseau," 99 (associating the Lisbon quake with an eruption of metaphysical debate in "popular literature" throughout Europe).

9. Charles Dickens, "Old Stories Retold: Earthquakes," *All the Year Round* 18 (1867): 545, 547, available at https://play.google.com/store/books/details?id=jXQHAQAAIAAJ.

10. Lucas Joel, "Benchmarks: November 1, 1755: Earthquake Destroys Lisbon," *Earth Magazine*, October 23, 2015, https://www.earthmagazine.org/article/benchmarks-november-1– 1755-earthquake-destroys-lisbon/; "Portugal Earthquake 1755," Portugal Visitor, https:// www.portugalvisitor.com/history/earthquake.

11. Gottfried Wilhelm Leibnitz, *Theodicy* (1710), ed. Austin M. Farrer, trans. E. M. Huggard (New York: Cosimo Classics, 2010), 138.

12. Alexander Pope, *An Essay on Man* (1734) (Teddington, UK: Echo Library, 2007), 9.

13. Dynes, "The Dialogue Between Voltaire and Rousseau," 112 (describing the view among social scientists that Rousseau's letter contains the "beginnings of a social science view of disaster" and noting that it prefigured current perspectives on disasters by two hundred years).

14. Dynes, "The Dialogue Between Voltaire and Rousseau," 106 (quoting Rousseau's letter).

15. Voltaire, *Candide* (1759), ed. Stanley Appelbaum (Mineola, NY: Dover, 1991), 87.

16. Dynes, "The Dialogue Between Voltaire and Rousseau," 106.

17. Bob Bolin, "Race, Class, Ethnicity, and Disaster Vulnerability," in *Handbook of Disaster Research*, ed. Havidán Rodrìguez, Enrico L. Quarantelli, and Russell R. Dynes (New York: Springer, 2007), 113, 114.

18. IPCC, "Annex I: Glossary," ed. J. B. R. Matthews, in Global Warming of 1.5°: An IPCC Special Report on the Impacts of Global Warming of 1.5°C above Pre-Industrial Levels and Related Global Greenhouse Gas Emission Pathways, in the Context of Strengthening the Global Response to the Threat of Climate Change, Sustainable Development, and Efforts to Eradicate Poverty, ed. V. Masson-Delmotte, P. Zhai, H.-O. Pörtner, D. Roberts, J. Skea, P. R. Shukla, A. Pirani, et al. (Geneva: United Nations, 2018).

19. Craig Zamuda, Daniel E. Bilello, Guenter Conzelmann, Ellen Mecray, Ann Satsangi, Vincent Tidwell, and Brian Walker, "Energy Supply, Delivery, and Demand," in Impacts, Risks, and Adaptation in the United States: Fourth National Climate Assessment, ed. D. R. Reidmiller, C. W. Avery, D. R. Easterling, K. E. Kunkel, K. L. M. Lewis, T. K. Maycock, and B. C. Stewart (Washington, DC: U.S. Global Change Research Program, 2018), 2:192.

20. See, for instance, Edward Hanna, Frank Pattyn, Francisco Navarro, Vincent Favier, Heiko Goelzer, Michiel R. van den Broeke, Miren Vizcaino, et al., "Mass Balance of the Ice Sheets and Glaciers—Progress Since AR5 and Challenges," Earth-Science Reviews 201 (2021): 102976, https://doi.org/10.1016/j.earscirev.2019.102976.

21. Voltaire, The Works of Voltaire (1764), 2:299, available at http://oll.libertyfund.org (quoting "Zoroaster," or Zarathustra).

22. "Working Group III: 10.4.2.2 Precautionary Considerations," Intergovernmental Panel on Climate Change, United Nations Framework Convention on Climate Change, art. 3.3, https://archive.ipcc.ch/ipccreports/tar/wg3/index.php?idp=437 ("adaptation measures").

23. Namely, the U.S. Forest Service, the Federal Emergency Management Agency, and the Department of Housing and Urban Development.

24. Blue Planet II, season 1, episode 5, "Green Seas," featuring David Attenborough, Peter Drost, and Roger Horrocks, aired November 26, 2017, https://www.bbc.co.uk/programmes /b09gl670.

25. Godfrey-Smith, Other Minds, 98.

26. Greta Thunberg, "'You Did Not Act in Time': Greta Thunberg's Full Speech to MPs," The Guardian, April 23, 2019https://www.theguardian.com/environment/2019/apr/23/greta -thunberg-full-speech-to-mps-you-did-not-act-in-time.

27. Godfrey-Smith, Other Minds, 64.

28. Godfrey-Smith, Other Minds, 75.

4. CLIMATE AND CASTE

This chapter is inspired by two previous articles of mine: "Diamond in the Rough: Pursuing Disaster Justice in Surat, India," Environment and Planning E: Nature and Space 1 (2018), and "Disaster Justice: The Geography of Human Capability," Duke Environmental Law & Policy Forum 24 (2012): 23.

1. Imani Brown and Samaneh Moafi, "Environmental Racism in Death Alley, Louisiana," Forensic Architecture, June 28, 2021, https://forensic-architecture.org/investigation/environ mental-racism-in-death-alley-louisiana.

2.　Tegan Wendland, "New Public Health Study Does Little to Allay Fears in Cancer Alley," WWNO (New Orleans Public Radio), March 4, 2021, https://www.wwno.org/coastal-desk /2021–03–04/new-public-health-study-does-little-to-allay-fears-in-cancer-alley; University Network for Human Rights, "Waiting to Die: Toxic Emissions and Disease Near the Louisiana Denkal/DuPont Plant," July 2019, 20 (identifying five census tracts in St. John the Baptist Parish, Louisiana, as having the highest cancer rates in the country); Kimberly Terrell and Gianna St. Julien, Tulane Law Clinic, "Toxic Air Pollution is Linked to Higher Cancer Rates among Impoverished Communities in Louisiana" (2021) (showing evidence of a statewide link between cancer rates and toxic air pollution in Louisiana and suggesting that toxic air pollution is a contributing factor to Louisiana's cancer burden), https:// law.tulane.edu/sites/law.tulane.edu/files/u1286/LTR%20Cancer%20Rates%20v%20Pollution -Related%20Risk%202021-6-21%20rev.%202021-6-23.pdf.

3.　Brown and Moafi, "Environmental Racism in Death Alley."

4.　United Nations, "Climate Justice," *Sustainable Development Goals*, May 31, 2019, https://www .un.org/sustainabledevelopment/blog/2019/05/climate-justice/.

5.　J. E. Haas, R. W. Kates, and M. J. Bowden, *Reconstruction Following Disaster: Major Insights* (Cambridge, MA: MIT Press, 1977), 176–77.

6.　Daniel A. Farber, James Ming Chen, Lisa Grow Sun, and Robert R. M. Verchick, *Disaster Law and Policy* (New York: Wolters Kluwer, 2015), 227–72.

7.　Christopher Flavelle, "Why Does Disaster Aid Often Favor White People?," *New York Times*, October 27, 2021, https://www.nytimes.com/2021/06/07/climate/FEMA-race-climate .html.

8.　Solomon Hsiang, Robert Kopp, and Trevor Houser, "Estimating Economic Damage from Climate Change in the United States," *Science* 356, no. 6345 (2017): 6–8, https://www .science.org/doi/10.1126/science.aal4369.

9.　Matthew T. Ballew, Adam R. Pearson, Jonathan P. Schuldt, John E. Kotcher, Edward W. Maibach, Seth A. Rosenthal, and Anthony Leiserowitz, "Is the Political Divide on Climate Change Narrower for People of Color? Evidence from a Decade of U.S. Polling," *Journal of Environmental Psychology*, 77, art. 101680 (Oct. 2021), https://doi.org/10.1016/j.jenvp .2021.101680.

10.　Robert R. M. Verchick, *Facing Catastrophe* (Cambridge, MA: Harvard University Press, 2010), 135.

11.　Aletta Brady, Anthony Torres, and Phillip Brown, "What the Queer Community Brings to the Fight for Climate Justice," *The Grist*, April 9, 2019, https://grist.org/article/what-the -queer-community-brings-to-the-fight-for-climate-justice/; Farber et al., *Disaster Law and Policy*; Jaimie Seaton, "Homeless Rates for LGTB Teens Are Alarming, But Parents Can Make a Difference," *Washington Post*, March 29, 2017.

12.　Alfred Tennyson, *In Memoriam A. H. H.*, Canto 56 (London: Edward Moxon, 1850).

13.　Gerald D. Berreman, "Caste in India and the United States," *American Journal of Sociology* 66, no. 2 (1960): 120–27, https://www.jstor.org/stable/2773155; Michelle Alexander, *The New Jim Crow: Mass Incarceration in the Age of Colorblindness* (New York: New Press, 2010), 21–22.

14. Isabel Wilkerson, *Caste: The Origins of Our Discontents* (New York: Random House, 2020), 70–71.

15. B. R. Ambedkar, *Annihilation of Caste with a Reply to Mahatma Gandhi* 43 (Chennai: Maven Books 2018), 43.

16. Wilkerson, *Caste*, 71, 384.

17. Emma Newburger, "Himalayan Glaciers Are Melting at an Extraordinary Rate, Research Finds," CNBC News, December 20, 2021, https://www.cnbc.com/2021/12/20/himalayan -glaciers-melting-at-extraordinary-rate-research-finds-.html; Kai Kresek, Marie Duraisami, "What's Happening in India's Forests?," *Global Forest Watch*, July 2, 2020, https://www .globalforestwatch.org/blog/data-and-research/whats-happening-in-india-forests/; Deepa Lakshmin, "The New Cool: Adopting Street Dogs Is Gaining Popularity in India," *National Geographic*, April 21, 2020, https://www.nationalgeographic.com/animals/article/street -dogs-indies-india-pets.

18. T. S. Eliot, "The Dry Salvages," in *Four Quartets* (New York: Harcourt 1943) 35, 35 ("I don't know much about gods; but I think that the river / Is a strong brown god.")

19. Prashasti Singh, "Fire at Hazira Gas Processing Plant in Surat Under Control: ONGC," *Hindustan Times*, September 24, 2020, https://www.hindustantimes.com/india-news /fire-breaks-out-at-ongc-plant-in-gujarat-s-surat/story-YykCkpVubh62dTFM8BJ8xI .html.

20. Bhavna Shah, Chirag Mistry, Alok J. Navik, and Ajay V. Shah, "Assessment of Heavy Metals in Sediments Near Hazira Industrial Zone at Tapti River Estuary, Surat, India," *Environmental Earth Sciences* 69 (2012): 12, https://www.researchgate.net/publication/233380183 _Assessment_of_heavy_metals_in_sediments_near_Hazira_industrial_zone_at_Tapti _River_estuary_Surat_India.

21. Amanda MacMillan and Jeff Turrentine, "Global Warming 101: How Is Global Warming Linked to Extreme Weather?," Natural Resource Defense Council, April 7, 2021, https:// www.nrdc.org/stories/global-warming-101.

22. Hermann E. Ott, Bernd Brouns, Wolfgang Sterk, and Bettina Wittneben, "It Takes Two to Tango—Climate Policy at COP 10 in Buenos Aires and Beyond," *Journal for European Environmental and Planning Law* 2 (2005): 84–91, 86.

23. "Indian Agriculture and Allied Industries Industry Report," India Brand Equity Foundation, November 2021, https://www.ibef.org/industry/agriculture-india.aspx.

24. Noah S. Diffenbaugh, Marshall Burke, "Global Warming Has Increased Global Economic Inequity," *Proceedings of the National Academy of Sciences of the United States of America*, April 22, 2019, https://www.pnas.org/content/116/20/9808.

25. See, generally, Jared Diamond, *Guns, Germs, and Steel: The Fate of Human Societies* (New York: Norton, 1997); Sambit Bhattacharyya, *Growth Miracles and Growth Debacles: Exploring Root Causes* (Cheltenham, UK: Edward Elgar, 2011); John Gallup and Jared Sachs, "Agriculture, Climate, and Technology: Why Are the Tropics Falling Behind?," *American Journal of Agriculture Economics* 82 (2000): 731–37; and Sambit Bhattacharyya, "The Historical Origins of Poverty in Developing Countries," in *The Oxford Handbook of the Social Science of Poverty*,

ed. David Brady and Linda M. Burton, https://www.oxfordhandbooks.com/view/10.1093
/oxfordhb/9780199914050.001.0001/oxfordhb-9780199914050-e-13?print=pdf.

26. Judith N. Shklar, *The Faces of Injustice* (New Haven, CT: Yale University Press, 1990) (developing the thesis presented in the Storrs Lectures, hosted at Yale Law School in 1988), 50.

27. Shklar, *The Faces of Injustice*, 50, 81.

28. Amartya Sen, "Equality of What?," Tanner Lecture on Human Values at Stanford University, May 22, 1979, https://www.ophi.org.uk/wp-content/uploads/Sen-1979_Equality-of
-What.pdf.

29. Amartya Sen, *Development as Freedom* (New York: Knopf Doubleday, 1999), 36, 70, 188;
Sen, "Equality of What?," 215–16.

30. Sen, *Development as Freedom*, 36.

31. Rise Module: Equality vs. Equity, The Interaction Institute for Social Change, https://
risetowin.org/what-we-do/educate/resource-module/equality-vs-equity/index.html.

32. Sen, *Development as Freedom*, 177–78.

33. Ambedkar, *Annihilation of Caste*.

34. Individual Assistance Program and Policy Guide (IAPPG), Version 1.1, FEMA (May 2021),
https://www.fema.gov/sites/default/files/documents/fema_iappg-1.1.pdf.

35. Infrastructure Investment and Jobs Act of 2021, 135 Stat. 429; Glen Thrush, "White House
Retrofits Infrastructure Bill to Better Help Poor Communities," *New York Times*, August 2,
2022, https://www.nytimes.com/2022/08/02/us/politics/biden-infrastructure-bill-poor
-communities.html; Inflation Reduction Act of 2022, P. Law 177-169.

36. CDC/ATSDR Social Vulnerability Index (SVI), Agency for Toxic Substances and Disease Registry, April 28, 2021, https://www.atsdr.cdc.gov/placeandhealth/svi/index.html.

37. Community Resilience Indicator Analysis: County-Level Analysis of Commonly Used
Indicators from Peer-Reviewed Research, Homeland Security, Argonne National Laboratory (2020), https://www.fema.gov/sites/default/files/2020-11/fema_community-resilience
-indicator-analysis.pdf. (The full list: Central Appalachian counties in Kentucky, West
Virginia, and Virginia; the Mississippi Delta region of Louisiana, Alabama, and Arkansas; southwestern Alabama and counties throughout the Southeast; counties in and tribal
nations in south and central South Dakota; counties in and tribal nations in New Mexico
and Arizona; South Texas; Puerto Rico; and the western coast and interior of Alaska.)

38. Mary Wollstonecraft, *A Vindication of the Rights of Woman*, ed. Carol H. Poston (Chicago:
University of Chicago Press, 1975), 71.

5. BELIEVING IS SEEING

1. Jennifer Carman, Matthew Goldberg, John Kotcher, Karine Lacroix, Anthony Leiserowitz, Edward Maibach, Jennifer Marlon, et al., "Global Warming's Six Americas, September 2021," Yale Program on Climate Change Communication, January 12, 2022, https://
climatecommunication.yale.edu/publications/global-warmings-six-americas-september
-2021/. The survey categorizes Americans into six different groups: alarmed (33 percent),

concerned (25 percent), cautious (17 percent), disengaged (5 percent), doubtful (10 percent), and dismissive (9 percent).

2. Matthew Ballew, Parrish Bergquist, John Kotcher, Anthony Leiserowitz, Edward Maibach, Jennifer Marlon, and Seth Rosenthal, "Which Racial/Ethnic Groups Care Most About Climate Change?," Yale Program on Climate Change Communication, April 16, 2020, https://climatecommunication.yale.edu/publications/race-and-climate-change/.

3. Lydia Saad, "Americans' Concerns About Global Warming on the Rise," Gallup Organization, April 8, 2013, http://www.gallup.com/poll/161645/americans-concerns-global -warming-rise.aspx (reporting on poll conducted March 7–10, 2013).

4. Jennifer Carman, Matthew Goldberg, Karine Lacroix, Anthony Leiserowitz, Edward Maibach, John Kotcher, Xinran Wang, and Seth Rosenthal, "Segmenting the Climate Change Alarmed: Active, Willing, and Inactive," Yale Program on Climate Change Communication, July 27, 2021, https://climatecommunication.yale.edu/publications/segmenting-the -climate-change-alarmed-active-willing-and-inactive/.

5. Dan M. Kahan, Donald Braman, Paul Slovic, John Gastil, and Geoffrey Cohen, "Cultural Cognition of the Risks and Benefits of Nanotechnology," *Nature Nanotechnology* 4 (February 2009): 87, https://www.nature.com/articles/nnano.2008.341.

6. Erica Dawson, Dan Kahan, Ellen Peters, and Paul Slovic, "Motivated Numeracy and Enlightened Self-Government," *Behavioral Public Policy* 1, 54–86 (September 2013): 1–22, https://dx.doi.org/10.2139/ssrn.2319992.

7. Chris Mooney, "Science Confirms: Politics Wrecks Your Ability to Do Math," *Mother Jones*, September 4, 2013, http://www.motherjones.com/politics/2013/09/new-study-politics -makes-you-innumerate.

8. Donald Braman, Dan M. Kahan, Gregory Mandel, Lisa Larrimore Ouellette, Ellen Peters, Paul Slovic, and Maggie Wittlin, "The Polarizing Impact of Science Literacy and Numeracy on Perceived Climate Change Risks," *Nature Climate Change* 2 (May 2012): 732–35, https://doi.org/10.1038/nclimate1547.

9. Donald Braman, Geoffrey L. Cohen, John Gastil, Dan. M. Kahan, and Paul Slovic, "Who Fears the HPV Vaccine, Who Doesn't, and Why? An Experimental Study of the Mechanism of Cultural Cognition," *Law & Human Behavior* 34, no. 6 (December 2010): 501–16, https://doi.org/10.1007/s10979-009-9201-0.

10. Brad Heath, "Americans Divided on Party Lines Over Risk from Coronavirus: Reuters/ Ipsos poll," *Reuters*, March 6, 2020, https://www.reuters.com/article/us-health-coronavirus -usa-polarization/americans-divided-on-party-lines-over-risk-from-coronavirus -reuters-ipsos-poll-idUSKBN20T2O3?fbcl (assessment of risk posed); "Does The Public Want to Get a COVID-19 Vaccine? When?," KFF, November 2021, https://www.kff.org /coronavirus-covid-19/dashboard/kff-covid-19-vaccine-monitor-dashboard/ (vaccination rates based on political affiliation); Xinyuan Ye, "Exploring the Relationship Between Political Partisanship and COVID-19 Vaccination Rate," *Journal of Public Health*, October 23, 2021, https://pubmed.ncbi.nlm.nih.gov/34693447/ (fatalities based on political affiliation).

11. *Handbook of Social Psychology*, ed. Susan T. Fiske, Daniel Todd Gilbert, and Gardner Lindzey, 2 vols. (Hoboken: Wiley, 2010), 2:801 (quoting Wilson).

12. Jonathan Haidt, "The Emotional Dog and Its Rational Tail: A Social Intuitionist Approach to Moral Judgment," *Psychological Review* 108 (2001): 814–34; see also Jonathan Haidt, *The Righteous Mind: Why Good People Are Divided by Politics and Religion* (New York: Penguin, 2012).

13. Aristotle, *Nicomachean Ethics*, trans. David Ross (Oxford: Oxford University Press, 2009), 2–4.

14. David Hume, *A Treatise of Human Nature*, in *The Essential Philosophical Works* (Ware, UK: Wordsworth Editions, 2011) 1, 360.

15. William J. Brennan Jr., "Reason, Passion, and the Progress of the Law," *Cardozo Law Review* 10, nos. 1–2 (October/November 1988): 3–24 (noting "the range of emotional and intuitive responses to a given set of facts or arguments, responses which often speed into our consciousness far ahead of the lumbering syllogisms of reason").

16. Brennan, "Progress of the Law," 3–24.

17. Katharine Hayhoe, *Saving Us: A Climate Scientist's Case for Hope and Healing in a Divided World* (New York: One Signal/Atria, 2021), 19.

18. K. K. Ottesen, "An Evangelical Scientist on Reconciling Her Religion and the Realities of Climate Change," *Washington Post*, March 2, 2021, https://www.washingtonpost.com /lifestyle/magazine/an-evangelical-scientist-on-reconciling-her-religion-and-the-realities -of-climate-change/2021/02/26/f757d1c2-40b7-11eb-8bc0-ae155bee4aff_story.html.

19. Ottesen, "Evangelical Scientist on Reconciling Her Religion."

20. Hayhoe, *Saving Us*, 18.

21. U.S. Department of Defense, Office of the Undersecretary for Policy (Strategy, Plans, and Capabilities), "Department of Defense Climate Risk Analysis," 2021, report submitted to the National Security Council, https://media.defense.gov/2021/oct/21/2002877353 /-1/-1/0/dod-climate-risk-analysis-final.pdf.

22. Donald Braman, Geoffrey L. Cohen, John Gastil, Dan M. Kahan, and Paul Slovic, "The Second National Risk and Culture Study: Making Sense of—and Making Progress in— The American Culture War of Fact," *GWU Legal Studies Research Paper No. 370* (October 2007): 23, http://ssrn.com/abstract=1017189 (describing influence of nuclear power on decision making); Donald Braman, Hank C. Jenkins-Smith, Dan M. Kahan, Carol L Silva, and Tor Tarantola, "Geoengineering and the Science Communication Environment: A Cross-Cultural Experiment," *Cultural Cognition Project Working Paper No. 92* (January 2012): 41, http://ssrn.com/abstract=1981907 (describing the influence of geoengineering on decision making).

23. Hayhoe, *Saving Us*, 90.

24. Hayhoe, *Saving Us*, 200.

25. For a more detailed discussion of this process, see Robert R. M. Verchick and Abby Hall, "Adapting to Climate Change while Planning for Disaster: Footholds, Rope Lines, and the Iowa Floods," *B.Y.U. Law Review* (2011): 2201.

26. Peter Boghossian and James Lindsay, *How to Have Impossible Conversations* (New York: Hachette Book Group, 2019), 272.

6. MOONSHOT ON THE BAYOU

1. Tom Yulsman, "On Eve of Oscar Bid by 'Beasts of Southern Wild,' New Report Finds Louisiana Sea Level Rise Fastest in World," *Discover*, February 23, 2013, https://www .discovermagazine.com/environment/on-eve-of-oscar-bid-by-beasts-of-southern-wild -new-report-finds-louisiana-sea-level-rise-fastest-in-world; Louisiana's Comprehensive Master Plan for a Sustainable Coast, *Coastal Protection and Restoration Authority* (2017), http://coastal.la.gov/wp-content/uploads/2017/04/2017-Coastal-Master-Plan_Web-Book _CFinal-with-Effective-Date-06092017.pdf (2,000 square miles of land loss).

2. Bob Marshall, Al Shaw, and Brian Jacobs, "Louisiana's Moon Shot to Rescue Its Coast," *Scientific American*, December 10, 2014, https://www.scientificamerican.com/article/louisiana -s-moon-shot-to-rescue-its-coast/.

3. Marshall et al., "Louisiana's Moon Shot to Rescue Its Coast."

4. W. V. Sweet, B. D. Hamlington, R. E. Kopp, C. P. Weaver, P. L. Barnard, et al., "Global and Regional Sea Level Rise Scenarios for the United States," National Oceanic and Atmospheric Administration (2022), https://aambpublicoceanservice.blob.core.windows .net/oceanserviceprod/hazards/sealevelrise/noaa-nos-techrpt01-global-regional-SLR -scenarios-US.pdf (sea level rise); Judith Kildow, Charles S. Colgan, Pat Johnston, Jason D. Scorse, and Maren Gardiner Farnum, "State of the U.S. Ocean and Coastal Economies: 2016 Update," National Ocean Economics Program (2016): 31, http://midatlanticocean.org /wp-content/uploads/2016/03/NOEP_National_Report_2016.pdf (Americans living near shorelines).

5. "The Coastline at Risk: 2016 Update to the Estimated Insured Value of U.S Coastal Properties," AIR Worldwide (2016), 10, https://www.air-worldwide.com/SiteAssets/Publications /White-Papers/documents/The-Coastline-at-Risk-2016 ($1 trillion in national wealth); E. Fleming, J. Payne, W. Sweet, M. Craghan, et al., "Impacts, Risks, and Adaptation in the United States: Fourth National Climate Assessment," *U.S. Global Change Research Program*, no. 2 (2018), 330, https://doi.org/10.7930/NCA4.2018.CH8 (half-a-trillion dollars' worth of real estate), 331 (99 percent of overseas trade).

6. Darryl Fears, "Built on Sinking Ground, Norfolk Tries to Hold Back Tide Amid Sea-Level Rise," *Washington Post*, June 17, 2012, https://www.washingtonpost.com/national/health -science/built-on-sinking-ground-norfolk-tries-to-hold-back-tide-amid-sea-level-rise /2012/06/17/gJQADUsxjV_story.html; "Norfolk Vision 2100," City of Norfolk (2016), 50, https://www.norfolk.gov/DocumentCenter/View/27768/Vision-2100--FINAL?bidId=.

7. Fleming et al., "Impacts," 322–52, 327, fig. 8.1.

8. Robert McCoppin, "Insurance Company Drops Suits Over Chicago-Area Flooding," *Chicago Tribune*, June 3, 2014, https://www.chicagotribune.com/news/breaking/chi-chicago -flooding-insurance-lawsuit-20140603-story.html (Farmers Insurance claim); *Jordan v.*

St. Johns County, 63 So. 3d 835 (Fla. Dist. Ct. App. 2011) (elevation of road); MaryAnn Spato, "Harvey Cedars Couple Receives $1 Settlement for Dune Blocking Ocean View," *NJ Advanced Media*, September 25, 2013, https://www.nj.com/ocean/2013/09/harvey_cedars _sand_dune_dispute_settled.html (protective sand dune).

9. Fleming et al., "Impacts," 440–41.

10. Thomas E. Dahl and Susan-Marie Stedman, *Status and Trends of Wetlands in the Coastal Watersheds of the Conterminous United States 2004 to 2009*, U.S. Department of the Interior, Fish and Wildlife Service, National Oceanic and Atmospheric Administration, and National Marine Fisheries Service (2013), 46, https://www.fws.gov/wetlands/documents /status-and-trends-of-wetlands-in-the-coastal-watersheds-of-the-conterminous-us-2004 -to-2009.pdf.

11. Dean Klinkenberg, "The 70 Million-Year-Old History of the Mississippi River, *Smithsonian Magazine*, September 2020, https://www.smithsonianmag.com/science-nature/geological -history-mississippi-river-180975509/.

12. Joshua Lewis and Henrik Ernstson, "Contesting the Coast: Ecosystems as Infrastructure in the Mississippi River Delta," *Progress in Planning* (2019): 1–30, 7, fig. 3, 10.1016/j. progress.2017.10.003.

13. Clarence H. Webb, "The Extent and Content of Poverty Point Culture," *American Antiquity* 33, no. 3 (1968): 300–302 https://doi.org/10.2307/278700; "Poverty Point World Heritage Site," Louisiana Office of Tourism (2022), https://www.povertypoint.us.

14. For a full and captivating account of the Great Mississippi River Flood, see John M. Barry, *Rising Tide: The Great Mississippi Flood of 1927 and How It Changed America* (New York: Simon & Schuster, 2007).

15. Samy Magdy, "Rising Seas Threaten Egypt's Fabled Port City of Alexandria," *AP News*, August 30, 2019, https://apnews.com/article/e4fec321109941798cdbefae310695aa (Alexandria); "'Great Distress': Bangladesh Bears Brutal Cost of Climate Crisis," *Aljazeera*, November 3, 2021, https://www.aljazeera.com/news/2021/11/3/bangladesh-global-warming -climate-crisis-seawater-agriculture (Bangladesh); Charles Schmidt, "New Elevation Measure Shows Climate Change Could Quickly Swamp the Mekong Delta," *Scientific American*, August 28, 2019, https://www.scientificamerican.com/article/new-elevation-measure -shows-climate-change-could-quickly-swamp-the-mekong-delta/.

16. Janet McConnaughey, "NASA Looks at Louisiana Delta System, Eyes Global Forecasts," *AP News*, June 29, 2021, https://apnews.com/article/louisiana-climate-change-science -technology-environment-and-nature-fd198096ea945b4c256564b841c1a336.

17. "Environmental Atlas of Lake Pontchartrain," U.S. Geological Survey (1998), https://pubs .usgs.gov/of/2002/ofo2-206/env-status/table.html#table5; "West Ponchartrain-Maurepas Swamp," *Audubon*, https://www.audubon.org/important-bird-areas/west-pontchartrain -maurepas-swamp.

18. Diane Austin, Bob Carriker, Tom McGuire, Joseph Pratt, et al., "History of the Offshore Oil and Gas Industry in Southern Louisiana," U.S. Department of the Interior, no. 1 (2004), https://www.lsu.edu/ces/publications/2004/2004-049_Final_Report.pdf.

19. Karen Savage, "Louisiana's Sinking Parishes Sue Fossil Fuel Companies Over Climate Change," *The Climate Docket*, October 10, 2017, https://www.climatedocket.com/2017/10/10 /louisiana-oil-gas-wetlands-climate-damage/.

20. "Offshore Oil and Gas: Updated Regulations Needed to Improve Pipeline Oversight and Decommissioning," U.S. Government Accountability Office (2021), https://www.gao.gov /products/gao-21-293.

21. For a study on how climate change considerations are assessed in Environmental Impact Statements under the National Environmental Policy Act, see Jessica A. Wentz, Grant Glovin, and Adrian Ang, "Survey of Climate Change Considerations in Federal Environmental Impact Statements, 2012–2014," Columbia Law School (2016), https://scholarship .law.columbia.edu/sabin_climate_change/15/.

22. Alyson Flournoy, William Andreen, et al., "Regulatory Blowout: How Regulatory Failures Made the BP Disaster Possible, and How the System Can Be Fixed to Avoid a Recurrence," Center for Progressive Reform (October 2010), https://cpr-assets.s3.amazonaws .com/documents/BP_Reg_Blowout_1007.pdf.

23. Mark Schleifstein, "New Orleans Files Wetland Damage Suit Against Oil, Gas Companies," *Advocate/Times Picayune* (New Orleans), April 1, 2019, https://www.nola.com/news /environment/article_601e0eaf-c33b-53c1-8872-6887c3c5cd90.html. For a vivid account of the first of these lawsuits, see Nathaniel Rich, "The Most Ambitious Environmental Lawsuit Ever," *New York Times*, October 2, 2014, https://www.nytimes.com/interactive/2014 /10/02/magazine/mag-oil-lawsuit.html.

24. See, e.g., John Schwartz, "A Mini-Mississippi River May Help Save Louisiana's Vanishing Coast," *New York Times*, February 25, 2020, https://www.nytimes.com/2020/02/25/climate /louisiana-mississippi-river-model.html; Elizabeth Kolbert, "Louisiana's Disappearing Coast," *New Yorker*, March 25, 2019, https://www.newyorker.com/magazine/2019/04/01 /louisianas-disappearing-coast.

25. CPRA, *Comprehensive Master Plan* (2017), ES-12 (seafood statistics); David Batker, Isabel de Torre, Robert Costanza, Paula Swedeen, John W. Day, et al., "Gaining Ground: Wetlands, Hurricanes, and the Economy: The Value of Restoring the Mississippi River Delta," Portland State University (2010), 7 (value of delta's "natural capital").

26. CPRA, *Comprehensive Master Plan* (2017): 14–15.

27. "About CPRA," Coastal Protection and Restoration Authority, https://coastal.la.gov /about/.

28. Craig E. Colton, *State of Disaster: A Historical Geography of Louisiana's Land Loss Crisis* (Baton Rouge: Louisiana State University Press, 2021): 38–39, 132–33.

29. PRA, *Comprehensive Master Plan* (2017), 96, fig. 4.2 (budget for structural defenses); Mark Schleifstein, "$3 Billion Morganza to the Gulf Levee to Get First $12.5 Million from Federal Government," *Advocate/Times-Picayune* (New Orleans), January 19, 2021, https://www .nola.com/news/business/article_f93d27c8-5a78-11eb-9f3c-d323db4e7635.html (Morganza-to-the-Gulf completion date).

30. CPRA, *Comprehensive Master Plan* (2017), 96, fig. 4.2.

31. "State of the Sector, Water Management 2016," *Greater New Orleans, Inc.*, https://gnoinc
.org/wp-content/uploads/sites/2/2020/07/state-of-the-sector-water-management.pdf.

32. CPRA, *Comprehensive Master Plan* (2017), 46.

33. James Hansen, Makiko Sato, Paul Hearty, Reto Ruedy, Maxwell Kelley, et al., "Ice Melt,
Sea Level Rise, and Superstorms: Evidence from Paleoclimate Data, Climate Modeling,
and Modern Observations That 2°C Global Warming Could Be Dangerous," *Copernicus
Publications* (2016), https://acp.copernicus.org/articles/16/3761/2016/acp-16-3761-2016.pdf.

34. CPRA, *Comprehensive Master Plan* (2017), 13–14.

35. CPRA, *Comprehensive Master Plan* (2017), 129–32.

36. Thomas McGarity, Sidney Shapiro, Karen Sokol, and David Flores, "Climate Justice, State
Courts and the Fight for Equity," Center for Progressive Reform (2019), https://cpr-assets
.s3.amazonaws.com/documents/Climate-Justice-FINAL-011219.pdf.

37. "Louisiana Coastal Wetlands Restoration Plan: Main Report and Environmental Impact
Statement," Louisiana Coastal Wetlands Conservation and Restoration Task Force (1993),
89, 119, https://www.doi.gov/sites/doi.gov/files/migrated/deepwaterhorizon/adminrecord
/upload/Louisiana-Coastal-Wetlands-Conservation-and-Restoration-Task-Force
-Louisiana-Coastal-Wetlands-Restoration-Plan-Main-Report-And-EIS-Nov-1993.pdf (sea
level rise is a part of "natural erosion processes"; on account of subsidence and global sea
level rise, bays are becoming "naturally deeper"; and finally, "subsidence and sea level rise
are natural processes that contribute to wetland deterioration and loss").

38. CPRA, *Louisiana's Comprehensive Master Plan for a Sustainable Coast*, 84 (2012): https://issuu
.com/coastalmasterplan/docs/coastal_master_plan-v2. Governor Jindal's skepticism—what
some then called "denial"—of the human contribution to the climate crisis has been examined by several journalists, including Ben Geman and National Journal, "Bobby Jindal's
Soft Climate-Change Skepticism," *The Atlantic*, September 16, 2014, https://www
.theatlantic.com/politics/archive/2014/09/bobby-jindals-soft-climate-change-skepticism
/446958/; Katie Glueck, "Jindal: White House 'Science Deniers'" *Politico*, September 16,
2014, https://www.politico.com/story/2014/09/bobby-jindal-white-house-science-deniers
-111003; and Ben Adler, "Meet the Climate Deniers Who Want to Be President," *Grist*,
August 20, 2014, https://grist.org/politics/meet-the-climate-deniers-who-want-to-be
-president/.

39. "Executive Order JBE 2020–18 of August 19, 2020, Climate Initiatives Task Force," (2020),
https://gov.louisiana.gov/assets/ExecutiveOrders/2020/JBE-2020–18-Climate-Initiatives
-Task-Force.pdf.

40. "Killer Red Fox," *National Geographic Presents*: IMPACT with Gal Gadot, Ep. 5., https://
www.youtube.com/watch?v=SF_5j0f0GHY.

41. "Killer Red Fox."

42. Maxine Burkett, Robert R. M. Verchick, and David Flores, "Reaching Higher Ground:
Avenues to Secure and Manage New Land for Communities Displaced by Climate
Change," Center for Progressive Reform, May 2017, https://cpr-assets.s3.amazonaws.com
/documents/ReachingHigherGround_1703.pdf.

43. Alaska Institute for Justice, "Rights of Indigenous People in Addressing Climate-Forced Displacement" (2020), http://climatecasechart.com/climate-change-litigation/wp-content/uploads/sites/16/non-us-case-documents/2020/20200116_NA_complaint.pdf.

44. Colton, *State of Disaster*, 152–53.

45. Jayur Madhusudan Mehta and Elizabeth L. Chamberlain, "Mound Construction and Site Selection in the Lafourche Subdelta of the Mississippi River Delta, Louisiana, USA," *Journal of Island and Coastal Archaeology* 14, no. 4 (2018), https://www.tandfonline.com/doi/abs/10.1080/15564894.2018.1458764?journalCode=uica20; Diana Yates, "Study: Ancient Mound Builders Carefully Timed Their Occupation of Coastal Louisiana Site," Illinois News Bureau, May 22, 2018, https://news.illinois.edu/view/6367/653233.

46. CPRA, *Comprehensive Master Plan* (2017), 30.

47. Nathaniel Rich, "Destroying a Way of Life to Save Louisiana," *New York Times Magazine*, July 21, 2020, https://www.nytimes.com/interactive/2020/07/21/magazine/louisiana-coast-engineering.html.

7. LIGHTS OUT

1. "Katrina Kills Most Fish in New Orleans Aquarium," CNN, September 7, 2005, http://edition.cnn.com/2005/TECH/science/09/07/katrina.zoos/ ("10,000 fish . . ."); Sheri Fink, *Five Days at Memorial: Life and Death in a Storm-Ravaged Hospital* (New York: Crown, 2013). Investigations showed that some patients had been injected with a combination of lethal drugs, that is, apparently euthanized by their caregivers. One doctor and two nurses were charged with second-degree murder, but a grand jury found no probable cause to indict them.

2. "Icebox Eye Sore Symbolizes Massive Cleanup Challenge," *USA Today*, October 20, 2005 (quoting Ron Fields).

3. Jessica Leigh Hester, "How Do This Year's Storms Stack Up Against Hurricane Sandy?," *Bloomberg CityLab*, October 31, 2017, https://www.bloomberg.com/news/articles/2017-10-31/how-does-hurricane-sandy-compare-to-harvey-irma-and-maria; Morgan McFall-Johnson, "Over 1,500 California Fires in the Past 6 Years—Including the Deadliest Ever—Were Caused by One," *Business Insider*, November 3, 2019, https://www.businessinsider.com/pge-caused-california-wildfires-safety-measures-2019-10. Pacific Gas & Electric, one of the nation's largest electric utilities, is now bankrupt and recently pleaded guilty to the deaths of 84 people killed in the 2018 Camp Fire. Ivan Penn and Peter Eavis, "PG&E Pleads Guilty to 84 Counts of Manslaughter in Camp Fire Case," *New York Times*, June 18, 2020, https://www.nytimes.com/2020/06/16/business/energy-environment/pge-camp-fire-california-wildfires.html.

4. For two excellent introductions to the power grid, see Gretchen Bakke, *The Grid: The Fraying Wires Between Americans and Our Energy Future* (New York: Bloomsbury, 2016); and Ted Koppel, *Lights Out: A Cyberattack, A Nation Unprepared, Surviving the Aftermath* (New York: Crown, 2015).

5. Michelle Davis and Steve Clemmer, *Power Failure: How Climate Change Puts Our Electricity at Risk* (Washington, DC: Union of Concerned Scientists, 2014), 4, http://www.ucsusa.org/global_warming/science_and_impacts/impacts/effects-of-climate-change-risks-on-our-electricity-system.html#.Vocus1d7VnY.

6. Davis and Clemmer, *Power Failure*, 6–7 (Laramie River Station; Duane Arnold nuclear plant); U.S. Government Accountability Office, *Climate Change: Energy Infrastructure Risks and Adaptation Efforts* (2014), 19–20, http://www.gao.gov/assets/670/660558.pdf (Brown's Ferry nuclear plant; Millstone Nuclear Power Station). For more on climate vulnerabilities and proposed solutions, see Sam C. A. Nierop and Michael B Gerrard, *Envisioning Resilient Electrical Infrastructure: A Policy Framework for Incorporating Future Climate Change into Electricity Sector Planning* (New York: Columbia Law School Center for Climate Change Law, 2013), http://web.law.columbia.edu/sites/default/files/microsites/climate-change/files/Publications/Students/envisioning_resilient_electrical_infrastructure.pdf.

7. Edgar Sandoval, Rick Rojas, and Allyson Waller, "Death Toll from Texas' Winter Storm Rises Sharply to 111," *New York Times,* March 25, 2021, https://www.nytimes.com/2021/03/25/us/texas-winter-storm-death-toll.html; David Roberts, "Lessons from the Texas Mess," *Volts*, February 24, 2021, https://www.volts.wtf/p/lessons-from-the-texas-mess; Eric Gimon, *Lessons from the Texas Big Freeze* (San Francisco: Energy Innovation, 2021), 4–5, https://energyinnovation.org/wp-content/uploads/2021/05/Lessons-from-the-Texas-Big-Freeze.pdf.

8. Erin Douglas and Ross Ramsey, "No, Frozen Wind Turbines Aren't the Main Culprit for Texas' Power Outages," *Texas Tribune*, February 16, 2021, https://www.texastribune.org/2021/02/16/texas-wind-turbines-frozen/ (quoting Sid Miller, Commissioner of Texas Department of Agriculture).

9. Glen sang about telephone lines, but still.

10. J. R. Minkel, "The 2003 Northeast Blackout—Five Years Later," *Scientific American*, August 13, 2008, https://www.scientificamerican.com/article/2003-blackout-five-years-later.

11. Chaamala Klinger, Owen Landeg, and Virginia Murray, "Power Outages, Extreme Events, and Health: A Systematic Review of the Literature from 2011–2012," *PLOS Currents Disasters*, January 2, 2014, http://currents.plos.org/disasters/article/power-outages-extreme-events-and-health-a-systematic-review-of-the-literature-from-2011-2012; Eric Klinenberg, "Adaptation: How Can Cities Be Climate-Proofed," *New Yorker*, December 30, 2012, http://www.newyorker.com/magazine/2013/01/07/adaptation-2 ("roughly seven times as many . . ."); Tracey Ross, *A Disaster in the Making: Addressing the Vulnerability of Low-Income Communities to Extreme Weather* (Washington, DC: American Progress, 2013), 10, https://www.americanprogress.org/wp-content/uploads/2013/08/LowIncomeResilience-2.pdf ("the annual air temperature . . ."); Bill M. Jesdale, Rachel Morello-Frosch, and Lara Cushing, "The Racial/Ethnic Distribution of Heat Risk-Related Land Cover in Relation to Residential Segregation," *Environmental Health Perspectives* 121 (2013): 812–13, https://ehp.niehs.nih.gov/doi/pdf/10.1289/ehp.1205919 (Berkeley survey).

12. Hurricane Sandy Rebuilding Task Force, *Hurricane Sandy Rebuilding Strategy: Stronger Communities, A Resilient Region* 14 (2013), https://portal.hud.gov/hudportal/documents /huddoc?id=hsrebuildingstrategy.pdf.

13. Lisa Friedman, "Trump Signs Order Rolling Back Environmental Rules on Infrastructure," *New York Times*, August 15, 2017, https://www.nytimes.com/2017/08/15/climate /flooding-infrastructure-climate-change-trump-obama.html?_r=0.

14. America's Wetland Foundation, America's Energy Coast, and Entergy, *Building a Resilient Energy Gulf Coast: Executive Report* (New Orleans: Entergy & America's Wetland Foundation, 2010), 6, http://www.entergy.com/content/our_community/environment/GulfCoast Adaptation/Building_a_Resilient_Gulf_Coast.pdf.

15. U.S. Energy Information Administration, "One in Three U.S. Households Faces a Challenge in Meeting Energy Needs," September 18, 2018, https://www.eia.gov/todayinenergy /detail.php?id=37072 (reporting on 2015 data); Basav Sen, Cynthia Bird, and Celia Bottger, *Energy Efficiency with Justice* (Washington, DC: Institute for Policy Studies, August 2018), 39, Fig. 23, https://ips-dc.org/report-energy-efficiency-with-justice/.

16. Con Edison, *Climate Change Resilience and Adaptation* (January 2021), 1, https://www.coned .com/-/media/files/coned/documents/our-energy-future/our-energy-projects/climate -change-resiliency-plan/climate-change-resilience-adaptation-2020.pdf.

17. Emma Foehringer Merchant, "What Superstorm Sandy Taught Consolidated Edison, 5 Years On," GreenTech Media, October 27, 2017, https://www.greentechmedia.com /articles/read/what-superstorm-sandy-taught-consolidated-edison.

18. Letter from Anne R. Siders, Columbia University Law School, to Jaclyn A. Brilling, N.Y. State Public Service Commission, December 12, 2012, available at https://web.law.columbia .edu/sites/default/files/microsites/climate-change/files/Publications/PSCPetitionNatura lHazardPlanning_0.pdf.

19. To review, a "one-hundred-year flood" is an event with a 1 percent chance of occurring in any given year. FEMA does not currently consider climate change when designating one-hundred-year-flood plains.

20. Karen Field, "Con Ed Nearing Completion of $1B Post-Sandy Fortification Effort," *T&D World*, November 21, 2017, https://www.tdworld.com/substations/article/20970523/con-ed -nearing-completion-of-1b-postsandy-fortification-effort.

21. Romany M. Webb, Michael Panfil, and Sarah Ladin, *Climate Risk in the Electricity Sector: Legal Obligations to Advance Climate Resilience Planning by Electric Utilities* (New York: Sabin Center for Climate Change Law, December 2020), 13, http://blogs.edf.org/climate411/files /2020/12/Climate-Risk-Electricity-Sector.pdf.

22. For more detail, see Con Edison, *Climate Change Resilience*.

23. New York Public Service Commission, Order Approving Electric, Gas and Steam Rate Plans in Accord with Joint Proposal (February 21, 2014), 72, https://perma.cc/Y78W -GY8H.

24. Webb, Panfil, and Ladin, *Climate Risk*, 16–26.

25. Davis and Clemmer, *Power Failure*, 9.

26. Davis and Clemmer, *Power Failure*, 12.

27. American Society of Civil Engineers, *Failure to Act: Closing the Infrastructure Investment Gap for America's Economic Future* (2017) 11, table 1, https://www.infrastructurereportcard.org/wp-content/uploads/2016/10/ASCE-Failure-to-Act-2016-FINAL.pdf.

28. See Infrastructure Investment and Jobs Act of 2021, 135 Stat. 429; Inflation Reduction Act of 2022, P. Law 117-169; Joan E. Greve, " 'Biggest Step Forward on Climate Ever': Biden Signs Democrats' Landmark Bill," *The Guardian*, August 16, 2022, https://www.theguardian.com/us-news/2022/aug/16/biden-signs-inflation-reduction-act-landmark-healthcare-climate-bill; Ben King, John Larsen, and Hannah Kolus, "A Congressional Climate Breakthrough," Rodium Group, July 28, 2022, https://rhg.com/research/inflation-reduction-act/ (projecting 40 percent reduction of carbon pollution below 2005 levels by 2030); Jesse D. Jenkins, Erin N. Mayfield, Jamil Farbes, Ryan Jones, Neha Patankar, Qingyu Xu, and Greg Schivley, "Preliminary Report: The Climate and Energy Impacts of the Inflation Reduction Act of 2022," Rapid Energy Policy Evaluation and Analysis Toolkit (Princeton University), August 2022, https://repeatproject.org/docs/REPEAT_IRA_Prelminary_Report_2022-08-04.pdf

29. For one take on performance-based metrics, see Dan Aas and Michael O'Boyle, *You Get What You Pay For: Moving Toward Value in Utility Compensation*, Part 2, 3 (San Francisco: America's Power Plan, June 2016).

30. Here I'm describing the city's oldest streetcar line, known as the "Green Line," which rumbles along St. Charles and Carrollton Avenues. There is also a "Red Line" (Canal Street-Cemeteries), an "Orange Line" (Canal Street-City Park/Museum), and a "Blue Line" (River Front).

31. Shalanda H. Baker, *Revolutionary Power: An Activist's Guide to the Energy Transition* (Washington, DC: Island Press, 2021). As my book goes to press, Baker serves in the Biden administration as deputy director for energy justice and secretary's advisor on equity at the U.S. Department of Energy. She is also a law professor (on leave) at Northeastern University.

32. Cecelia Bolon, Talia Lanckton, and Shalanda Baker, *Utilities 101: A Guide to the Basics of the Electric Utility Industry with a Focus on Justice* (Initiative for Energy Justice, 2020), 5, https://iejusa.org/wp-content/uploads/2020/08/Utilities-101-Full-Guide-v3.pdf (citing data from the U.S. Department of Energy); Ariel Drehobl and Lauren Ross, *Lifting the High Energy Burden in America's Largest Cities: How Energy Efficiency Can Improve Low Income and Underserved Communities* (Washington, DC: American Council for an Energy-Efficient Economy, April 2016), 21, fig. 4, https://www.aceee.org/sites/default/files/publications/researchreports/u1602.pdf.

33. For more on the interplay between energy affordability and race, see Editorial, "Energy Justice towards Racial Justice," *Nature* (August 2020), 5:551, https://www.nature.com/articles/s41560-020-00681-w.pdf.

34. For more detail, see Bolon, Lanckton, and Baker, *Utilities 101*, 12.

35. Tyler Mauldin, "Hurricane Laura Ravages Power Grid," CNN, August 30, 2020, https://www.cnn.com/2020/08/30/weather/laura-louisiana-power-lines-wx/index.html.

8. FLASH! CRACK! BOOM!

1. Edmond S. Meany, *Origin of Washington Geographic Names* (Seattle: University of Washington Press, 1923), 322.
2. Andrew Wineke, "Smoky Siege," Washington State Department of Ecology, September 22, 2020. https://ecology.wa.gov/Blog/Posts/September-2020/A-smoky-siege.
3. Wineke, "Smoky Siege."
4. Tim Gruver, "Wildfires Are Still Burning in the Pacific Northwest as Fire Restrictions Lifted on Federal Public Lands in Washington," *The Center Square*, September 30, 2020, https://www.thecentersquare.com/oregon/wildfires-are-still-burning-in-the-pacific-northwest-as-fire-restrictions-lifted-on-federal-public/article_2cb0a80c-0365-11eb-bc7f-cb6bf169ob44.html.
5. Tim Gruver, "'It's Like Detroit Has Disappeared': Oregon Wildfire Survivors Look Back on the Help That Never Came," *The Center Square*, February 16, 2021, https://www.thecentersquare.com/oregon/its-like-detroit-has-disappeared-oregon-wildfire-survivors-look-back-on-help-that-never-came/article_4dffd8a8-64f3-11eb-8239-0ba643758307.html.
6. "Into the Fire: A Smokejumper's Life," Frontline Wildfire Defense System, https://www.frontlinewildfire.com/into-fire-smokejumpers-life/.
7. Heather Hansman, "A Quiet Rise in Wildland Firefighter Suicides," *The Atlantic*, October 29, 2017, https://www.theatlantic.com/health/archive/2017/10/wildland-firefighter-suicide/544298/.
8. See "Total Wildland Fires and Acres (1960–2015)," National Interagency Fire Center, https://www.nifc.gov/fireInfo/fireInfo_stats_totalFires.html (reporting increasing rates of acreage burned by American wildfires); see also "Climate Change Indicators: Wildfires," U.S. Environmental Protection Agency, https://www.epa.gov/climate-indicators/climate-change-indicators-wildfires ("The extent of area burned by wildfires each year appears to have increased since the 1980s").
9. James M. Vose and David L. Peterson, "Forests," in *Fourth National Climate Assessment 2018*, ed. Greg Marland (Washington, DC: U.S. Global Change Research Program), 232–67, at 234.
10. Harmeet Kaur, "California Fire Is Now a 'Gigafire,' a Rare Designation for a Blaze That Burns at Least a Million Acres," CNN, October 6, 2020, https://www.cnn.com/2020/10/06/us/gigafire-california-august-complex-trnd/index.html.
11. Vose and Peterson, "Forests," 239.
12. Vose and Peterson, "Forests," 239, Fig. 6.3.
13. See U.S. Congress, House, Committee on Financial Services, "TRIA at Ten Years: The Future of the Terrorism Risk Insurance Program: Hearing Before the House

Subcommittee. on Insurance, Housing and Community Opportunity," 112th Cong., 2nd sess., 2012, 63 & no.2 (statement of Robert P. Hartwig, President & Economist, Insurance Info. Inst.) (discussing claim payouts by insurers to policyholders interchangeably with the term "insured losses"); Stefan Doerr, Cristina Santin, "Wildfire: a Burning Issue for Insurers?," Lloyd's (2013), 20 (reporting $7.9 billion in insured losses during the period from 2002 to 2011 and $1.7 billion in insured losses during the preceding ten-year period).

14. See John W. Schoen, "Cost of Western Blazes Spreads Like Wildfire," *NBC News*, August 22, 2013, http://www.nbcnews.com/business/cost-western-blazes-spreads-wildfire-6C109 74725; "Historical Census of Housing Tables: Home Values," U.S. Census Bureau, Housing and Household Economic Statistics Division, June 6, 2012, https://perma.cc/ 3UGU -DM2D.

15. See "Wildfire Funding Fix: What It Means for America's Forests," National Association of State Foresters, https://www.stateforesters.org/wp-content/uploads/2018/12/Wildfire -Funding-Fix-ONE-PAGER.pdf.

16. "The Rising Cost of Wildfire Operations: Effects on the Forest Service's Non-Fire Work," U.S. Forest Service, August 4, 2015, https://www.fs.usda.gov/sites/default/files/2015-Fire -Budget-Report.pdf.

17. Michael R. Darling, "A Baptism by Incentives: Curing Wildfire Law at the Front of Oil and Gas Regulation," *Texas Law Review* 96, no. 6 (2018): 1237.

18. Tim Gruver, "It's Like Detroit Has Disappeared: Oregon Wildfire Survivors Look Back on Help That Never Came," *The Center Square*, February 16, 2021, https://www.thecen tersquare.com/oregon/its-like-detroit-has-disappeared-oregon-wildfire-survivors-look -back-on-help-that-never-came/article_4dffd8a8-64f3-11eb-8239-0ba643758307.html.

19. Sarah Zielinski, "What Do Animals Do in a Wildfire?," *National Geographic*, July 22, 2014, https://www.nationalgeographic.com/animals/article/140721-animals-wildlife-wildfires -nation-forests-science.

20. "Australia's Fires 'Killed or Harmed Three Billion Animals,'" BBC News, July 29, 2020, https://www.bbc.com/news/world-australia-53549936.

21. Dino Grandoni, "The Energy 202: California's Fires Are Putting a Huge Amount of Carbon Dioxide into the Air," *Washington Post*, September 17, 2020, https://www.washing tonpost.com/politics/2020/09/17/energy-202-california-fires-are-putting-huge-amount -carbon-dioxide-into-air/.

22. California Climate Policy Dashboard," *Berkeley Law*, https://www.law.berkeley.edu /research/clee/research/climate/climate-policy-dashboard/.

23. "Ground-level Ozone Pollution: Ground-level Ozone Basics," U.S. Environmental Protection Agency, https://www.epa.gov/ground-level-ozone-pollution/ground-level-ozone-basics #effects.

24. "Particulate Matter (PM) Pollution: Particulate Matter (PM) Basics," U.S. Environmental Protection Agency, https://www.epa.gov/pm-pollution/particulate-matter-pm-basics.

25. "About Underlying Cause of Death, 1999–2019," Centers for Disease Control and Prevention, *CDC Wonder*, 2018, https://wonder.cdc.gov/ucd-icd10.html; Jay S. Coggins, Andrew L. Goodkind, Jason D. Hill, and Christopher W. Tessum, "Fine-Scale Damage Estimates of Particular Matter Air Pollution Reveal Opportunities for Location-Specific Mitigation of Emissions," *Proceedings of the National Academy of Sciences* 116, no. 18 (April 30, 2019): 8775–80, https://www.pnas.org/content/116/18/8775#ref-27.

26. Christopher G. Nolte, "Air Quality," in *Fourth National Climate Assessment*, ed. David D'Onofrio (Washington, DC: U.S. Global Change Research Program, 2018), 514.

27. E. A. Burakowski, J. L. Campbell, N. J. Casson, A. R. Contosta, M. S. Crandall, I. F. Creed, M. C. Eimers, et al., "Winter Weather Whiplash: Impacts of Meteorological Events Misaligned with Natural and Human Systems in Seasonally Snow-Covered Regions," *Earth's Future* 7, no. 12 (November 2019): 1434–50, https://doi.org/10.1029/2019EF001224.

28. Nolte, "Air Quality," 512.

29. Jennifer K. Balch, Bethany A. Bradley, John T. Abatzoglou, R. Chelsea Nagy, Emily J. Fusco, and Adam L. Mahood, "Human-Started Wildfires Expand the Fire Niche Across the United States," *Proceedings of the National Academy of Sciences* 114, no. 11 (March 14, 2017): 2948.

30. Daniel L. Swain, "Attributing Extreme Events to Climate Change: A New Frontier in a Warming World," *One Earth* 2, no. 6 (June 19, 2020): 526, figure 4, https://www.cell.com/one-earth/fulltext/S2590-3322(20)30247-5.

31. Geert J. Van Oldenborgh, Folmer Krikken, Sophie Lewis, Nicholas J. Leach, Flavio Lehner, Kate R. Saunders, Michiel van Weel, et al., "Attribution of the Australian Bushfire Risk to Anthropogenic Climate Change," *National Hazards and Earth System Sciences* 21 (2021): 951, https://www.worldweatherattribution.org/wp-content/uploads/WWA-attribution_bushfires-March2020.pdf.

32. Megan Kirchmeier-Young, Nathan Gillett, Francis Zwiers, Alex J. Cannon, and Faron Anslow, "Attribution of the Influence of Human Induced Climate Change on an Extreme Fire Season," *Earth's Future* 7, no. 1 (December 13, 2018): 2, https://agupubs.onlinelibrary.wiley.com/doi/full/10.1029/2018EF001050.

33. Richard Seager, Allison Hooks, A. Park Williams, Benjamin Cook, Jennifer Nakamura, and Naomi Henderson, "Climatology, Variability, and Trends in the United States Vapor Pressure Deficit, an Important Fire-Related Meteorological Quantity," *Journal of Applied Meteorology and Climatology* 54, no. 6 (July 2014): 3, http://ocp.ldeo.columbia.edu/res/div/ocp/WestCLIM/PDFS/Seager_etal_VPD.pdf.

34. A. Park Williams, John T. Abatzoglou, Alexander Gershunov, Janin Guzman-Morales, Daniel A. Bishop, Jennifer K. Balch, and Dennis P. Lettenmaier, "Study Bolsters Case that Climate Change Is Driving Many California Wildfires," AGU, July 15, 2019, https://news.agu.org/press-release/study-bolsters-case-that-climate-change-is-driving-many-california-wildfires/.

35. Diana Leonard and Andrew Freedman, "Western Wildfires: An 'Unprecedented,' Climate Change Fueled Event, Experts Say," *Washington Post*, September 11, 2020, https://www.washingtonpost.com/weather/2020/09/11/western-wildfires-climate-change/.

36. "The 1910 Fires," *Forest History Society*, https://foresthistory.org/research-explore/us-forest-service-history/policy-and-law/fire-u-s-forest-service/famous-fires/the-1910-fires/.

37. Umair Irfan, "California Has 129 Million Dead Trees. That's a Huge Wildfire Risk," *Vox*, September 4, 2018, https://www.vox.com/2018/9/1/17800358/california-mendocino-wildfire-dead-trees.

38. Ryan Sabalow, "Trump Calls for More Logging on Federal Lands to Fight Wildfires," *Sacramento Bee*, December 21, 2018, https://www.sacbee.com/news/politics-government/article223445505.html; "Bush Unveils 'Healthy Forests' Plan," CNN, August 22, 2002, http://www.cnn.com/2002/ALLPOLITICS/08/22/bush.timber/.

39. Ted Sickinger, "Would More Logging Have Averted Oregon's Catastrophic 2020 Wildfires? Probably Not," *Oregonian*, February 7, 2021, https://www.oregonlive.com/wildfires/2020/10/oregons-labor-day-wildfires-raise-controversial-questions-about-how-forests-are-managed.html.

40. Sickinger, "Would More Logging Have Averted Oregon's Catastrophic 2020 Wildfires?"

41. Eliza Barclay, David Roberts, and Umair Irfan, "California's Recurring Wildfire Problem, Explained," *Vox*, September 10, 2020, https://www.vox.com/21430638/california-wildfires-2020-orange-sky-august-complex.

42. Sandra B. Zellmer and Jan Laitos, "Only the Russian Federation, Brazil, and Canada Have More Forests Than the United States," *Principles of Natural Resources Law* (St. Paul: West Academic Publishing, 2014), 230.

43. Sickinger, "Would More Logging Have Averted Oregon's Catastrophic 2020 Wildfires?"

44. Tania Schoennagel, "Adapt to More Wildfire in Western North American Forests as Climate Changes," *Proceedings of the National Academy of Sciences* 114, no. 18 (May 2, 2017): 4586, https://doi.org/10.1073/pnas.1617464114.

45. Schoennagel, "Adapt to More Wildfire," 4586.

46. Kale Williams, "Controlled Burns in Oregon: Can More Fires Create Less Smoke?," *The Oregonian*, August 29, 2019, https://www.oregonlive.com/news/erry-2018/10/9eefddfd8c9177/controlled-burns-in-oregon-can.html; Shannon Gormley, "Oregon's Indigenous Communities Know how to Stop Megafires. Will the State Let Them?," *Willamette Week*, October 7, 2020, https://www.wweek.com/news/2020/10/07/oregons-indigenous-communities-know-how-to-stop-megafires-will-the-state-let-them/.

47. Williams, "Controlled Burns in Oregon."

48. Sickinger, "Would More Logging Have Averted Oregon's Catastrophic 2020 Wildfires?"

49. Williams, "Controlled Burns in Oregon."

50. Stephen M. Strader, "Spatiotemporal Changes in Conterminous US Wildfire Exposure from 1940 to 2010," *Natural Hazards* (February 15, 2018): 2, 6, 1, https://docs.wixstatic.com/ugd/fc36a1_9415de656b0445b09ba1022078f8a6d5.pdf.

51. Tim Ellis, "In Areas at High Risk for Wildfires, Relative Affordability Lures Homebuyers," *Redfin News*, October 16, 2020, https://www.redfin.com/news/home-prices-rise-slower-wildfire-risk/.

52. Ellis, "Relative Affordability Lures Homebuyers."

53. Julia Falcon, "Homes in High-Risk Wildfire Areas Are More Affordable," *Housing Wire*, October 16, 2020, https://www.housingwire.com/articles/homes-in-high-risk-wildfire-areas-are-more-affordable/.

54. "Wildfire Risk to Communities," U.S. Department of Agriculture: Forest Service, https://wildfirerisk.org/wp-content/uploads/2020/04/WRC-Info-Sheet-2020-04.pdf.

55. U.S. Department of Agriculture: Forest Service, "Wildfire Risk to Communities."

56. "FAQ," U.S. Department of Agriculture: Forest Service, https://wildfirerisk.org/about/faq/.

57. "Paradise, Butte County, California," U.S. Department of Agriculture: Forest Service, https://wildfirerisk.org/explore/3/06/06007/0600055520/.

58. Edvard Munch, MM T 2367, 1892, paper, Munch Museum, Oslo, https://emunch.no/TRANS_HYBRIDMM_T2367.xhtml.

59. Fred Prata, Alan Robock, and Richard Hamblyn, "The Sky in Edvard Munch's *The Scream*," *Journal of the American Meteorological Society* 99, no. 7 (July 1, 2018): 1380, 1382, https://journals.ametsoc.org/view/journals/bams/99/7/bams-d-17-0144.1.xml.

60. Edvard Munch, *eMunch.No: Text and Image*, ed. Mai-Britt Guleng (Oslo: Munch Museum, 2011), 276.

61. Timothy Ingalsbee, "Getting Burned: A Taxpayer's Guide to Wildfire Suppression Costs," *Firefighters United for Safety, Ethics, & Ecology* (August 2010): 4.

62. Daniel A. Farber, James Ming Chen, Robert R. M. Verchick, and Lisa Grow Sun, *Disaster Law and Policy* (Dordrecht: Wolters Kluwer 2015), 53.

63. See, for instance, "Wildland-Urban Interface," California Department of Housing and Community Development, https://www.hcd.ca.gov/building-standards/state-housing-law/wildland-urban-interface.shtml.

64. Isaac Gendler, Isaac, "How Urban Growth Boundary Could Impact Los Angeles," Abundant Housing LA, August 10, 2020, https://abundanthousingla.org/how-an-urban-growth-boundary-could-impact-los-angeles/.

65. Daniel Cusick, "A Rebuilt Paradise Nervously Watches Wildfire on the Horizon," *Scientific American*, September 14, 2020, https://www.scientificamerican.com/article/a-rebuilt-paradise-nervously-watches-wildfire-on-the-horizon/.

66. Firefighters United for Safety, Ethics & Ecology, https://fusee.org/home/policy-advocacy; National Wildfire Coordinating Group, https://www.nwcg.gov/; California Wildfire Relief Fund, https://latinocf.org/norcal-wildfire-relief-fund/; Save Our Forests, https://keeporegongreen.org/contribute/.

9. YUCCAS, GARDENERS, AND ZOOKEEPERS

1. Some say the Joshua tree served as a model for the ill-starred Truffula tree in Dr. Seuss's *The Lorax* (New York: Random House, 1971), but I have my doubts. A better match seems to be the Monterey cypress, a rare tree native to the California coast. See Jennifer Billock, "Visit the Original Lorax Tree in Dr. Seuss's San Diego," *Smithsonian.com*, August 5, 2016, https://www.smithsonianmag.com/travel/visit-seussical-san-diego-180959997/.

2. Scott Martelle, "See Those Joshua Trees While You Can. Climate Change Is Killing Our National Parks," *Los Angeles Times*, September 25, 2018, https://www.latimes.com/opinion/la-ol-enter-the-fray-climate-change-has-heavier-impact-on-1537892055-htmlstory.html.

3. My discussion of the Joshua tree here draws from James W. Cornett, *The Joshua Tree*, 2nd ed. (Palm Springs, CA: Nature Trails Press, 2018), 13; Joseph W. Zarki, *Joshua Tree National Park* (Mount Pleasant, ME: Arcadia Publishing, 2015); Kenneth L. Cole, Kirsten Ironside, Jon Eischeild, Gregg Garfin, Phillip B. Duffy, and Chris Toney, "Past and Ongoing Shifts in Joshua Tree Distribution Support Future Modeled Range Contraction," *Ecological Applications* 21, no. 1 (2011): 137–49; and Cameron W. Barrows and Michelle L. Murphy-Mariscal, "Modeling Impacts of Climate Change on Joshua Trees at Their Southern Boundary: How Scale Impacts Predictions," *Biological Conservation* 152 (2012): 29–36.

4. Diana Wells, *The Lives of Trees: An Uncommon History* (Chapel Hill, NC: Algonquin Books of Chapel Hill, 2010), 180 (quoting John C. Fremont); see also Cornett, *The Joshua Tree*, 14.

5. Cornett, *The Joshua Tree*, 14 (quoting Chase and Larson).

6. My discussion of Minerva Hoyt draws from Lary M. Dilsaver, *Preserving the Desert: A History of Joshua Tree National Park* (Staunton, VA: George F. Thompson Publishing, 2017); Tracy Conrad, "How Minerva Hamilton Saved Joshua Tree Park for Future Generations," *Desert Sun*, May 14, 2019, https://www.desertsun.com/story/life/2019/05/11/how-minerva-hamilton-hoyt-saved-joshua-tree-national-park/1179516001/; Deborah Netburn, "How a Southern California Matron Used Her Wits and Wealth to Create Joshua Tree National Park," *Los Angeles Times*, February 18, 2019, https://www.latimes.com/science/sciencenow/la-sci-col1-joshua-tree-minerva-hoyt-20190214-htmlstory .html. To my knowledge, there is no book-length biography of Minerva Hoyt, an obvious gap in the preservationist literature.

7. Netburn, "A Southern California Matron" (quoting Holt).

8. *Desert Protection Act of 1994*, Public Law 103–433, *U.S. Statutes at Large* 410 §2 (1994): 4471–72.

9. David N. Cole, Constance I. Millar, and Nathan L. Stephenson, "Responding to Climate Change: A Toolbox of Management Strategies," in *Beyond Naturalness: Rethinking Park and Wilderness Stewardship in an Era of Rapid Change,* ed. David N. Cole and Laurie Yung (Washington, DC: Island Press, 2010), 184.

10. The amount of refugia depends on the success of future greenhouse-gas reductions. See Lynn Sweet, Tyler Green, James G. C. Heintz, Neil Frakes, Nicolas Graver, Jeff S.

Rangitsch, Jane E. Rodgers, Scott Heacox, and Cameron W. Barrows, "Congruence Between Future Distribution Models and Empirical Data for an Iconic Species at Joshua Tree National Park," *Ecosphere* 10, no. 6 (2019): 1.

11. See Philip Kiefer, "Iconic Joshua Trees May Disappear, But Scientists Are Fighting Back." *National Geographic*, October 15, 2018, https://www.nationalgeographic.com/environment /2018/10/joshua-trees-moths-threatened-climate-change-scientists-seek-solutions/; Jennifer Harrower and Gregory S. Gilbert, "Context-Dependent Mutualisms in the Joshua Tree–Yucca Moth System Shift Along a Climate Gradient," *Ecosphere* 9, no. 9 (2018).

12. Elizabeth Shogren, "How the National Park Service Is Planning for Climate Change," *The Atlantic*, August 24, 2016, http://www.theatlantic.com/science/archive/2016/08/how-the -national-park-service-is-planning-for-climate-change/497042/ (quoting Jarvis).

13. Elizabeth Shockman, "Climate Change Is a Huge Threat to Our National Parks," *PRI*, June 19, 2016, https://www.pri.org/stories/2016-06-19/climate-change-huge-threat-our -national-parks (quoting Jarvis).

14. See Kellie Lunney, "Parks Backlog Bill Has $6.4B Price Tag Over Next Decade," *Greenwire*, September 30, 2019, https://www.eenews.net/greenwire/stories/1061184047

15. William B. Monahan and Nicholas A. Fisichelli, "Climate Exposure of US National Parks in a New Era of Change," *PLOS ONE* 9, no. 7 (2014): 9, https://doi.org/10.1371/journal.pone .0101302.

16. E. M. Leibensperger, L. J. Mickley, D. J. Jacob, W. T. Chen, J. H. Seinfeld, A. Nenes, P. J. Adams. D. G. Streets, N. Kumar, and D. Rind, "Climatic Effects of 1950–2050 Changes in U.S. Anthropogenic Aerosols—Part 2: Climate Response," *Atmospheric Chemistry and Physics* 12, no. 7 (2012): 3349–62, https://doi.org/10.5194/acp-12-3349-2012.

17. See, generally, Maria A. Caffrey, Rebecca L. Beavers, and Cat Hawkins Hoffman, *Sea Level Rise and Storm Surge Projections for the National Park Service* (Fort Collins, CO: National Park Service, 2018), https://www.nps.gov/subjects/climatechange/upload/2018-NPS-Sea-Level -Change-Storm-Surge-Report-508Compliant.pdf.

18. See Alejandro E. Camacho and Robert L. Glicksman, "Legal Adaptive Capacity: How Program Goals and Processes Shape Federal Land Adaptation to Climate Change," *University of Colorado Law Review* 87 (2016): 714.

19. See, generally, Caffrey et al., *Sea Level Rise*.

20. U.S. Environmental Protection Agency, "Climate Impacts in the Southwest," *EPA*, January 19, 2017, https://19january2017snapshot.epa.gov/climate-impacts/climate-impacts -southwest_.html.

21. See Cole et al., "Past and Ongoing Shifts."

22. Cole et al., "Past and Ongoing Shifts," 143.

23. This warming would culminate in what scientists call the "Mid-Holocene Warm Period," dating from seven thousand to five thousand years ago, in which temperatures in many parts of the world were warmer than they are today. National Oceanic and Atmospheric Administration, "Mid-Holocene Warm Period," https://www.ncdc.noaa.gov/global-warming /mid-holocene-warm-period (last visited October 20, 2019).

24. Edward Abbey, *Desert Solitaire: A Season in the Wilderness* (New York: Ballantine Books, 1971).

25. See Nicolas Brulliard, "The Famous Landmark the Park Service Almost Encased in Plastic," National Parks Conservation Association, September 26, 2017, https://www.npca.org /articles/1646-the-famous-landmark-the-park-service-almost-encased-in-plastic.

26. Rebecca Morelle, "Mariana Trench: Deepest-Ever Sub Dive Finds Plastic Bag," BBC News, May 13, 2019, https://www.bbc.com/news/science-environment-48230157.

27. See C. S. Holling, ed., *Adaptive Environmental Assessment and Management* (London: John Wiley and Sons, 1978).

28. See David L. Peterson, Constance I. Millar, Linda A. Joyce, Michael J. Furniss, Jessica E. Halofsky, Ronald P. Neilson, and Toni Lyn Morelli, *Responding to Climate Change on National Forests: A Guidebook for Developing Adaptation Options* (Portland, OR: U.S. Department of Agriculture, Forest Service, Pacific Northwest Research Station, 2011), 76, Box 21.

29. Lee E. Frelich and Peter B. Reich, "Wilderness Conservation in an Era of Global Warming and Invasive Species: A Case Study from Minnesota's Boundary Waters Canoe Area Wilderness," *Natural Areas Journal* 29 (2009): 390–91.

30. Elisabeth Long and Eric Biber, "The Wilderness Act and Climate Change Adaptation," *Environmental Law* 44 (2014): 658.

31. M. Martin Smith and Fiona Gow, "Unnatural Preservation," *High Country News*, February 4, 2008, https://www.hcn.org/issues/363/17481/print_view (quoting Graber).

32. James Burchfield and Martin Nie, *National Forests Policy Assessment Report to Montana Senator Jon Tester* (Missoula: University of Montana, 2008), 11.

33. Camacho and Glicksman, "Legal Adaptive Capacity," 744–45.

34. *Perkins v. Bergland*, 608 F.2d 803, 806 (9th Cir. 1976). Despite what some skin-care advertisers say, pores don't breathe.

35. 16 U.S.C. §§ 1131–1136 (2006 and Supp. II 2008), amended by Pub. L. No. 111–11 (2006).

36. J. B. Ruhl, "Climate Change Adaptation and the Structural Transformation of Environmental Law." *Environmental Law* 40, no. 2 (2010): 393–94, https://law.lclark.edu/live/files/5649.

37. William C. Tweed, "An Idea In Trouble: Thoughts About the Future of Traditional National Parks in the United States," *National Park Service Centennial Essay Series* 27, no. 1 (2010): 6, 8, http://www.georgewright.org/271tweed.pdf.

38. Exec. Order No. 13653 (2013). This order was rescinded by President Trump in 2017. See Exec. Order No. 13783 (2017).

39. Camacho and Glicksman, "Legal Adaptive Capacity," 758.

40. National Forest Management Act, 16 U.S.C. § 1603 (2012).

41. *California Desert Protection Act of 1994*, Public Law 103–433, *U.S. Statutes at Large* 410 §2 (1994): 4471–72.

42. Camacho and Glicksman, "Legal Adaptive Capacity," 819.

43. See, generally, Eric Biber and Elisabeth Long Esposito, "The National Park Service Organic Act and Climate Change," *Natural Resources Journal* 56 (2016); Long and Biber, "The Wilderness Act."

44. Long and Biber, "The Wilderness Act," 627.

45. See K. M. Archie, L. Dilling, J. N. Milford, and F. C. Pampel, "Climate Change and West-ern Public Lands: A Survey of U.S. Federal Land Managers on the Status of Adaptation Efforts" *Ecology and Society* 17, no. 24 (2012): 20, http://dx.doi.org/10.5751/ES-05187-170420 (West Mountain survey); L. C. Jantarasami, J. J. Lawler, and C. W. Thomas, "Institutional Barriers to Climate Change Adaptation in U.S. National Parks and Forests," *Ecology and Society* 15 (2010): 33 (Washington State survey).

46. See, for instance, L. C. Jantarasami, R. Novak, R. Delgado, E. Marino, S. McNeeley, C. Narducci, J. Raymond-Yakoubian, L. Singletary, and K. Powys Whyte, "Tribes and Indig-enous Peoples," in *Impacts, Risks, and Adaptation in the United States: Fourth National Climate Assessment, Volume II*, ed. D. R. Reidmiller, C. W. Avery, D. R. Easterling, K. E. Kunkel, K. L. M. Lewis, T. K. Maycock, and B. C. Stewart (Washington, DC: U.S. Global Change Research Program, 2018), 572–603.

47. *Massachusetts v. U.S. Environmental Protection Agency*, 549 U.S. 497 (2007).

10. THE OCTOPUS'S GARDEN

1. Sylvia A. Earle, *Sea Change: A Message of the Oceans* (College Station: Texas A&M Univer-sity Press, 2021), xxv.

2. See Jules Verne, *Twenty Thousand Leagues under the Seas; or The Marvelous and Exciting Adventures of Pierre Arronnax, Conseil His Servant, and Ned Land, a Canadian Harpooner*, trans. F. P. Walter (Orinda, CA: Seawolf Press, 2018) (1872). It's a favorite adventure novel of mine, but you shouldn't rely on it for accurate science. The reference to leagues, by the way, represents a horizontal measurement. The book's narrator, marine biologist Pierre Arronnax, calculates that his marvelous and exciting voyage abord the Nautilus, a technologically advanced submarine, extended twenty thousand leagues, or forty thou-sand nautical miles—twice the distance of the circumference of the Earth.

3. Important titles on the shelf would include Deborah Rowan Wright, *Future Sea: How to Rescue and Protect the World's Oceans* (Chicago: University of Chicago Press, 2022); Robin Kundis Craig, *Comparative Ocean Governance: Place-Based Protections in an Era of Climate Change* (Cheltenham, UK: Edward Elgar, 2012), and *Climate Change and Ocean Governance: Politics and Policy for Threatened Seas*, ed. Paul G. Harris (Cambridge: Cambridge Univer-sity Press, 2019).

4. Coral Restoration Foundation, https://www.coralrestoration.org.

5. Aristotle, *Historia Animalium* VIII (4–10), 588b ("We are at a loss to know whether [corals] are animals or plants"); James Bowen, *The Coral Reef Era: From Discovery to Decline* (New York: Springer, 2015), 4 (presuming Aristotle believed corals to be "rocks").

6. "Earth from Space: Great Barrier Reef," European Space Agency, March 4, 2009, https://www.esa.int/Applications/Observing_the_Earth/Earth_from_Space_Great_Barrier_Reef.

7. Melanie McField, Nadia Bood, Ana Fonseca, Alejandro Arrivillaga, et al., "Status of the Mesoamerican Reef After the 2005 Coral Bleaching Event," National Oceanic and

Atmospheric Administration, https://www.coris.noaa.gov/activities/caribbean_rpt/SCRBH 2005_05.pdf.

8. See the Ocean Portal Team and Jennifer Bennett (NOAA), "Ocean Acidification: Zooplankton," *Smithsonian Ocean*, March 9, 2022, https://ocean.si.edu/ocean-life/inverte brates/ocean-acidification.

9. For more on the exotic world of "microbiota," see "The Microbiota," Microbiome Foundation, March 9, 2022, https://microbiome-foundation.org/the-microbiota/?lang=en.

10. David Souter, Serge Planes, Jérémy Wicquart, Murray Logan, et al., "Status of Coral Reefs of the World: 2020," *Global Coral Reef Monitoring Network* (2020): 3, https://gcrmn.net/wp -content/uploads/2021/10/Executive-Summary-with-Forewords.pdf, 1.

11. Souter et al., "Status of Coral Reefs."

12. Souter et al., "Status of Coral Reefs," 3, 19.

13. Hawthorne L. Beyer, Emma V. Kennedy, Maria Beger, Chaolun Allen Chen, et al., "Risk-Sensitive Planning for Conserving Coral Reefs under Rapid Climate Change," *Conservation Letters*, ed. Amanda Lombard (2018), 2, https://conbio.onlinelibrary.wiley.com/doi/full /10.1111/conl.12587.

14. Neil Vigdor, "Sprawling Coral Reef Resembling Roses Is Discovered off Tahiti," *New York Times*, January 20, 2022.

15. "Global Coral Reef Habitat Maps," *Allen Coral Atlas*, https://allencoralatlas.org.

16. *County of Maui, Hawaii v. Hawaii Wildlife Fund*, 140 S.Ct. 1462 (2020).

17. "The Harsh Truth About Coral Reefs in Maui," *Kai Kanani*, January 2, 2020, https://www .kaikanani.com/harsh-truth-coral-reefs-in-maui/.

18. 33 U.S.C.S. § 1342.

19. Transcript of Oral Argument, *County of Maui, Hawaii v. Hawaii Wildlife Fund* (18–260), Oyez, https://www.oyez.org/cases/2019/18-260.

20. *County of Maui*, 140 S.Ct. at 1477.

21. Edward Goodwin, *International Environmental Law and the Conservation of Coral Reefs* (Oxford: Routledge, 2011), 49-50.

22. Enric Sala and Sylvaine Giakoumi, "No-Take Marine Reserves Are the Most Effective Protected Areas in the Ocean," *ICES Journal of Marine Science* 75, no. 3 (2017): 1166–68, https:// academic.oup.com/icesjms/article/75/3/1166/4098821.

23. "Why the Famed Ocean Advocate Sylvia Earle Is Looking South," *Antarctica 2020*, https:// antarctica2020.org/why-famed-ocean-advocate-sylvia-earle-is-looking-south/.

24. Jane Lubchenco and Kirsten Grorud-Colvert, "Making Waves: The Science and Politics of Ocean Protection," *Science* 350 (2015): 6259, 10.1126/science.aad5443.

25. Alexa Hargreaves-Allen, Susana Mourato, and Eleanor Jane Milner-Gulland, "Drivers of Coral Reef Marine Protected Area Performance," *PLoS ONE* 12, no. 6 (2017): e0179394, https://doi.org/10.1371/journal.pone.0179394.

26. "Why the Famed Ocean Advocate Sylvia Earle Is Looking South."

27. Lisa Boström-Einarsson, Russell C. Babcock, Elisa Bayraktarov, Daniela Ceccarelli, Nathan Cook, Sebastian C. A. Ferse, Boze Hancock, et al., "Coral Restoration—A

Systematic Review of Current Methods, Successes, Failures, and Future Directions," *PLoS ONE* 15, no. 1 (2020): e0226631, https://doi.org/10.1371/journal.pone.0226631.

28. Erica Cirino, "Coral in Crisis: Can Replanting Efforts Stop Reefs' Death Spiral?," *The Revelator*, January 24, 2020, https://therevelator.org/coral-reef-replanting/ (quoting Anderson).

29. Beyer et al., "Risk-Sensitive Planning," 4.

30. Catrin Einhorn and Christopher Flavelle, "A Race Against Time to Rescue a Reef from Climate Change," *New York Times*, December 5, 2020.

31. Einhorn and Flavelle, "A Race Against Time."

32. DiveN2Life, http://www.diven2life.org.

33. See Kama Lee Cannon and Marsha L. Carn, "SCUBA Diving: Motivating and Mentoring Culturally and Cognitively Diverse Adolescent Girls to Engage in Place-Based Science Enrichment," *The Education Forum* 84, no. 1 (2019): 71, https://doi.org/10.1080/00131725.2019.1649508.

34. Katharine Hayhoe, *Saving Us: A Climate Scientist's Case for Hope and Healing in a Divided World* (New York: One Signal/Atria, 2021), 224.

11. THE LONG GOODBYE

This chapter is inspired by a previous article of mine: "The Long Goodbye: How to Build a Responsible Climate Migration Program," *Temple Law Review* 93, no. 4 (Summer 2021): 713–734.

1. The Jean Charles Choctaw Nation was previously known as the "Isle de Jean Charles Band of Biloxi-Chitimacha-Choctaw Tribe." For more on the community's long and complicated efforts to relocate, see Tristan Baurick, "The Last Days of Isle de Jean Charles: A Louisiana Tribe's Struggle to Escape the Rising Sea," *Advocate/Times Picayune* (New Orleans), August 28, 2022; U.S. Government Accountability Office (GAO), "Climate Change: A Climate Migration Pilot Program Could Enhance the Nation's Resilience and Reduce Federal Fiscal Exposure," GAO-20-488 (July 6, 2020), https://www.gao.gov/products/gao-20-488. I draw extensively on this document throughout this chapter. See also "About the Isle De Jean Charles Resettlement," Louisiana Office of Community Development (2021), https://isledejeancharles.la.gov/about-isle-de-jean-charles-resettlement; Coral Davenport and Campbell Robertson, "Resettling the First American 'Climate Refugees,' " *New York Times*, May 2, 2016; Ted Jackson, "On the Louisiana Coast, a Native Community Sinks Slowly into the Sea," *Yale Environment* 360, March 15, 2018, http://e360.yale.edu/features/on-louisiana-coast-a-native-community-sinks-slowly-into-the-sea-isle-de-jean-charles. For a closeup look, see my website, http://www.robverchick.com.

2. Adam Terando, Lynne Carter, Kirstin Dow, Kevin Hiers, Kenneth E. Kunkel, et al., *Impacts, Risks, and Adaptation in the United States: Fourth National Climate Assessment 2, Southeast* (2018): 743, 761, https://nca2018.globalchange.gov/downloads/NCA4_Ch19_Southeast_ExecSum.pdf.

3. Baurick, "The Last Days of Isle de Jean Charles"; Julie Dermansky, "Isle de Jean Charles Tribe Turns Down Funds to Relocate First US 'Climate Refugees' as Louisiana Buys Land

Anyway," *DeSmog*, January 11, 2019, http://www.desmogblog.com/2019/01/11/isle-de-jean -charles-tribe-turns-down-funds-relocate-climate-refugees-louisiana; Julie Dermansky, "Louisiana and Isle de Jean Charles Tribe Seek to Resolve Differing Visions for Resettling 'Climate Refugees,'" *DeSmog*, February 5, 2019, https://www.desmogblog.com/2019 /02/05/louisiana-isle-de-jean-charles-tribe-plans-resettlement-climate-refugees.

4. Christopher Flavelle, "The Toughest Question in Climate Change: Who Gets Saved?," *Bloomberg*, August 29, 2016, http://www.bloomberg.com/opinion/articles/2016-08-29/the -toughest-question-in-climate-change-who-gets-saved; Rachel Waldholz, "Obama Denies Newtok's Request for Disaster Declaration," Alaska Public Media, January 18, 2017, https://www.alaskapublic.org/2017/01/18/obama-denies-newtoks-request-for-disaster -declaration/.

5. Rachel Waldholz, "Obama Denies Newtok's Request"; Greg Kim, "With Virus Funds, Newtok Will Build More Homes in Mertarvik," Alaska Public Media, July 30, 2020, http:// www.alaskapublic.org/2020/07/30/with-boon-of-funding-newtok-faces-questions-of -how-to-best-get-remaining-residents-to-mertarvik/.

6. Mathew E. Hauer, "Migration Induced by Sea-Level Rise Could Reshape the US Population Landscape," *Nature Climate Change* 7, no. 321, 2017; Maxine Burkett, David Flores, and Robert Verchick, "Reaching Higher Ground: Avenues to Secure and Manage New Land for Communities Displaced by Climate Change," Center for Progressive Reform (2017): 6–7, https://cpr-assets.s3.amazonaws.com/documents/ReachingHigherGround _1703.pdf.

7. Anne Barnard, "The $119 Billion Sea Wall that Could Defend New York . . . Or Not," *New York Times*, January 17, 2020; Hilary Whiteman, "Staten Island Seawall: Designing for Climate Change," CNN Style, July 14, 2019, https://www.cnn.com/style/article/staten -island-seawall-climate-crisis-design/index.html.

8. Burkett, Flores, and Verchick, "Reaching Higher Ground," 6–7 (displaying two maps showing the communities in the contiguous United States as well as Alaska that are in the process of climate migration).

9. Alexa Jay, Kristin Lewis, David Reidmiller, and Katie Reeves, "The Fourth National Climate Assessment and Beyond: Communicating the Science and Risks of Climate Change," paper presented at the Geological Society of America Annual Meeting, Seattle, January 2017, http://dx.doi.org/10.1130/abs/2017AM-305621.

10. White House Office of the Press Secretary, "Fact Sheet: President Obama Announces New Investments to Combat Climate Change and Assist Remote Alaskan Communities," press release, September 2, 2015, http://obamawhitehouse.archives.gov/the-press-office/2015/09 /02/fact-sheet-president-obama-announces-new-investments-combat-climate; Christopher Flavelle, "Obama's Final Push to Adapt to Climate Change," *Bloomberg Opinion*, December 16, 2016, https://www.bloomberg.com/opinion/articles/2016-12-16/obama-s-final-push -to-adapt-to-climate-change.

11. Executive Order 14,008 of January 27, 2021, Tackling the Climate Crisis at Home and Abroad, Federal Registry, vol. 86, no. 19 (2021): 7619–23, 7626, 7629.

12. Davenport and Robertson, "Resettling the First American 'Climate Refugees,'"; Convention Relating to the Status of Refugees, July 28, 1951, 19 U.S.T. 6259, 189 U.N.T.S. 150 (defining "refugee" under international law); Camila Ruz, "The Battle over the Words Used to Describe Migrants," BBC, August 28, 2015, https://www.bbc.com/news/magazine -34061097.

13. Lindsey Jacobson, "In Areas Hit Hard by Climate Change, Only the Rich Can Afford to Stay," CNBC, September 16, 2016, https://www.cnbc.com/2021/09/16/climate-change-sea -level-rise-devalue-some-homes.html.

14. Robert R. M. Verchick and Lynsey Rae Johnson, "When Retreat Is the Best Option: Flood Insurance After Biggert-Waters and Other Climate Change Puzzles," *John Marshall Law Review* 47, no. 2 (2013): 707.

15. Diane P. Horn and Baird Webel, "Introduction to the National Flood Insurance Program (NFIP)," Congressional Research Service (2021): 27, https://sgp.fas.org/crs/homesec /R44593.pdf.

16. Elizabeth Fleming, Michael Craghan, John Haines, Juliette Finzi Hart, Heidi Stille, and Ariana Sutton-Griem, "Coastal Effects," in Terando et al., *Impacts, Risks, and Adaptation in the United States*, 330, https://nca2018.globalchange.gov/chapter/8/; Verchick and Johnson, "When Retreat Is the Best Option," 717.

17. John Gramlich and Alissa Scheller, "What's Happening at the U.S.-Mexico Border in 7 Charts," Pew Research Center, November 9, 2021, https://www.pewresearch.org/fact-tank /2021/11/09/whats-happening-at-the-u-s-mexico-border-in-7-charts/.

18. Nisha Agarwal, Kayly Ober, T. Alexander Aleinikoff, Maria Otero, J. Brian Atwood, Anne C. Richard, Reuben Brigety, et al., *Task Force Report to the President on the Climate Crisis and Global Migration*, Refugees International, July 14, 2021, https://www.refugee sinternational.org/reports/2021/7/12/task-force-report-to-the-president-on-the-climate -crisis-and-global-migration-a-pathway-to-protection-for-people-on-the-move.

19. Dermansky, "Differing Visions."

20. Jeremy Martinich, Benjamin DeAngelo, Delavane Diaz, Brenda Ekwurzel, Guido Franco, Carla Frisch, James McFarland, and Brian O'Neill, *Reducing Risks Through Emissions Migration*, in *Fourth National Climate Assessment*, 1346, 1372–73.

21. Mark Schleifstein, "No Levees in Corps' $1.8 Billion Flood Plan for Southwest Louisiana," *Times-Picayune*, March 21, 2015.

22. Anne Stauffer, Justin Theal, and Colin Foard, *Natural Disaster Mitigation Spending Not Comprehensively Tracked*, Pew Charitable Trusts (September 2018): https://www.pewtrusts .org/-/media/assets/2018/09/fiscal_federalism_federal_and_state_funding_issue_brief _v1.pdf.

23. Stauffer, Theal, and Foard, *Natural Disaster Mitigation*; Pew Charitable Trusts, *What We Don't Know About State Spending on Natural Disasters Could Cost Us* (Washington, DC: Pew Charitable Trusts, 2018), 2.

24. Robert R. M. Verchick and Abby Hall, "Adapting to Climate Change While Planning for Disaster: Footholds, Rope Lines, and the Iowa Floods," *Brigham Young University Law*

Review 2011, no. 6: 2223–30; Robert R. M. Verchick, "Disaster Justice: The Geography of Human Capability," *Duke Environmental Law and Policy Forum* 23 (2012): 67–68.

25. Nathan Rott, "Biden's Climate Change Plans Could Face Serious Challenges in a Divided Congress," NPR, November 10, 2020, http://www.npr.org/2020/11/10/933548716/bidens -climate-change-plans-could-face-serious-challenges-in-a-divided-congress.

12. PERSIST AND PREVAIL

1. Katharine Hayhoe, *Saving Us: A Climate Scientist's Case for Hope and Healing in a Divided World* (New York: One Signal Publishers/Atria, 2021), 213.

2. Hayhoe, *Saving Us*, 232.

Index

www.ingramcontent.com/pod-product-compliance
Ingram Content Group UK Ltd.
Pitfield, Milton Keynes, MK11 3LW, UK
UKHW041949130225
455076UK00003B/71

9 780231 219013